Restoring the Pillars of Life

Other titles available from Éditions La Butineuse:

Farmers have the Earth in Their Hands, Paul Luu

Land and Climate. Insights from the IPCC Special report, Patrick Love

Energy chronicles. Keys to understanding the importance of energy, Greg de Temmerman

Feeding the Earth. A Manifesto for Regenerative Agriculture, Daniel Baertschi

Farmers have the Earth in Their Hands, Paul Luu, with Marie-Christine Bidault

The Climate General, Tom Middendorp, with Antonie van Campen

Cover and interior design: © Agence Coam

ISBN: 978-2-493291-75-2

© 2021 Éditions La Butineuse
Atelier des Entreprises
Place de l'Europe – Porte Océane 3
56400 Auray – France
www.editions-labutineuse.com

Ananda Fitzsimmons

Restoring the Pillars of Life

ÉDITIONS la Butineuse

Table of contents

Introduction

Life on Earth is changing. Our climate is changing. We live in a time where the average temperature rises with each decade, a process of warming that is accelerating. According to the State of Global Climate report, published by the World Meteorological Organization, 2023 was the hottest year on record, with the global average near-surface temperature at 1.45° Celsius above the pre-industrial baseline. It was the warmest ten-year period on record. For the temperature to remain stable, a balance must be struck between the amount of heat from the Sun that is absorbed by the Earth's surface and atmosphere and the amount reflected back into space. Clearly the Earth now retains more heat than it releases. We know that greenhouse gasses in the atmosphere trap heat. When we talk about climate change, most people think of the greenhouse effect.

But what if the problem is more complex? What if, in addition to trapping more heat in Earth's atmosphere, we are also destroying the infrastructure that regulates temperature? Although burning less fossil fuel is absolutely critical, achieving reduction targets will take long and are not sufficient in themselves to avoid reaching irreversible tipping points. Our modern lifestyle has removed much of the natural infrastructure used to sustain the carbon and water cycles, two of the Earth's most important climate regulating systems.

Not only is the Earth getting warmer, we are living in the time of the sixth mass extinction of life on the planet. The number of species of animals, plants and insects going extinct is accelerating. Since 1970, we have lost 70% of our species biodiversity. This biodiversity is a big part of the infrastructure which regulates climate. **We are at fault when we talk about the climate crisis and the biodiversity crisis as if they were two separate crises. Healthy diverse ecosystems are key to regulating and restoring climate.**

In *Hydrate the Earth*, I gave the example of John D Liu who documented a massive ecosystem restoration project on the Loess Plateau in China. In one area of the plateau, little more than a patch of barren desert, restoration was achieved without planting anything. Workers dug trenches in a grid pattern. Afterwards they piled bales of straw into the trenches. Then they waited for rain, and when in the desert it does rain, it usually rains heavily in a short time. The result was that microbes, which were dormant in the wet straw, began to create soil. They needed carbon and water to activate. They got to work decomposing the straw, releasing nutrients, which in turn activated other small creatures and dormant seeds. Vegetation began to grow, further breaking down the straw. This new vegetation acted as a seal, keeping water so that it didn't evaporate. A cascade of life spread slowly through the desert. Every spurt of life attracted more life, more biodiversity. As the land became more hospitable, more plants grew, more birds and insects arrived in a widening and virtuous circle of life.

John D Liu's story shows how carbon, water and biodiversity are the three pillars of life. But wherever modern industrial society establishes itself a systematic decline of carbon in the soil, a reduction in the quantity and quality of water resources and a steady decline in species biodiversity inevitably follows.

The purpose of this book is to show how these three pillars of life are interconnected, and understand the fundamental role they play in preserving life on our blue planet.

And see how we all can contribute to stopping a climate disaster and re-establish a flourishing life on Earth. Healthy natural ecosystems play a crucial role not only in creating resilience to the effects of climate change but they can actually help to mitigate it.

The IPCC clearly states that ecosystem restoration must go along side reducing our greenhouse gas footprint, yet not enough people are focused on this. When I believed that the only thing that could be done about climate change was reducing the use of fossil fuels, I felt only governments and big corporations could avert the climate crisis. I could cut my energy use or sign petitions, but I doubted the real impact of those decisions. When it became clear that other contributing factors were involved, I saw many more avenues for change. We are going to hear the stories of many inspiring people who found their path to making a difference by working with the pillars of life.

I hope they will empower you to find ways to be part of the solution. I have included some experiential learning exercises to help you understand the concepts concretely, not only as abstractions. They can be done in a school classroom, in any community-organised learning group, or just on your own.

Chapter One
Life Cycles

Almost all energy on Earth comes from the Sun in the form of heat and light. Natural cycles distribute this energy, transform it and spread it around the world. Carbon and water are two of the primary drivers of the balance of heat on our planet as they change from gas to liquid to solid using the Sun's energy.

Water is the primary buffer of our climate. Without water vapour and other greenhouse gases in the atmosphere, the Sun's heat would quickly radiate back into space, leaving the Earth a cold barren place. Carbon, once sequestered in solid form, remains out of the atmosphere for a time. Each time carbon or water change phase using the Sun's energy, they redistribute heat. When the carbon and water cycles are balanced, life on Earth thrives.

Carbon, water and biodiversity are the three pillars of life. The first two – carbon and water – are the main building blocks on which the third depends. If we want to continue to have a liveable planet, we have no choice but to understand better how these cycles work, how human activity has disrupted them, and how we can help to restore the balance.

Create your own mini water cycle in a bottle.

Find a large jar with a tight-fitting lid. Cut out a piece of fine mesh screen to fit its diameter. Find a few medium sized stones to form the bottom layer. Then place the screen over the stones. Now find a place among some mosses and small low growing plants. With a small shovel dig up the soil at least to the root depth of the plants. Place a layer of soil (worms, bugs and all) over your screen. The screen keeps the soil from falling to the bottom of the jar. Then place some small plants, cover them with a bit of moss, but leave enough room above them so they can grow. Water your little garden and screw on the lid.

Place the jar in a location where there is Sun for at least part of the day and watch the water cycle in action.

You will never have to open the jar to water the plants. Water will settle to the bottom around the stones. The plants will absorb the water, then, through their leaves, release it into the air as water vapour (this process is called 'evapotranspiration'). You will see this vapour condense on the sides of the jar and run back down. Your little terrarium will keep on recycling its own water, carbon and oxygen without any help from you!

Closed Loop Systems: The Biosphere 2 Experiment

A sustainable system is a closed loop system. It feeds energy into itself to perpetuate continuous recycling over a long time. For example, dead vegetation becomes the food for the next generation of plants, helping them to grow. Ecosystems have evolved in such a way that species complement each other's life cycle. A bird follows a herd of ruminant animals to feed off the insect larva which hatch in the manure of the grazers. The wastes of the two species fertilise the soil, which enables the grasses to grow lush. Grasses nourish the ruminants and the cycle repeats. Nothing is wasted. How different is this from modern society in which we generate mountains of waste, with less and less room to dispose of it?

In the early 90's a gigantic research facility was constructed by a group in Oracle, Arizona. The group was influenced by Buckminster Fuller's concept of Spaceship Earth, the notion that we are passengers on this planet, hurtling through space with only a limited number of resources. The group wanted to create a sustainable closed loop system. Their project was also motivated by the desire to find solutions to surviving a potential nuclear disaster and to establishing future colonies on other planets. The project was called Biosphere 2; Biosphere 1 was planet Earth.

Biosphere 2 was a closed facility which included areas designed as miniature versions of natural biomes: a rainforest, an ocean with a coral reef, a mangrove wetland, a savannah and a fog desert. The biomes were stocked with the insects, birds and fish normally present in such ecologies. Members of the community had living quarters, laboratories and workshops. They grew their own food and raised domestic animals, including pigmy goats, chickens, dwarf pigs and tilapia fish. They also treated their own waste material. Eight "biospherians" (as they called themselves) chose to live in this alternative world bubble for two years, attracting great interest and debate from the media.

The debate continues as to whether the experiment succeeded. Although they did not fully succeed in creating the perfectly balanced closed loop biosphere they hoped for, there was no accumulation of toxins within the facility, and the farm was among the world's highest producing per surface area. All the biospherians were in excellent health when they came out.

Yet they did report being hungry much of the time and found living together not altogether easy. The biggest issue, which required outside intervention, was that oxygen levels gradually declined. After sixteen months they had fallen to such dangerously low levels that they showed symptoms of oxygen deprivation. After many months of trying to adjust the balance of carbon and oxygen in the biosphere, it was discovered that the cause was some exposed concrete in the structure, which was absorbing oxygen. Outside air needed to be pumped in. There were ecological imbalances as well. Pollinator species did not survive; there were outbreaks of invasive insects; the so-called natural biomes failed to maintain their ecological balance; condensation made the desert too wet; morning glories overtook the rainforest, blocking the light. The team struggled constantly to compensate these imbalances.

All of the urine and faeces generated, all of the water used in washing, all of the food waste and waste material from the animals had to be sanitised and recycled in a natural closed loop system that didn't pollute. One of the biospherians, Mark Nelson, was in charge of waste water systems at Biosphere 2. He went on from his time there to create more similar systems all over the world, that he described in his book *The Wastewater Gardener: Preserving the Planet One Flush at a Time*. The whole experience was an education in understanding the biological complexities of the closed loop cycles of life in a way that modern society has painfully neglected.

It was, however, a profound learning experience for everyone involved. Imagine having to think about everything you use in your life, where it comes from, where it goes? What is involved in producing your food, your clothing, your house, your car, and what happens when you no longer have any use for it?

The Carbon Cycle

Carbon in its gas form is CO_2: one molecule of carbon bound to two of oxygen. Carbon is severed from its two oxygen molecules through photosynthesis. Using the Sun's energy, plants breathe in CO_2 and breathe out oxygen. Only green plants – and phytoplankton - have this ability. It is a real "superpower". Without it, carbon would remain in its gas form.

A plant draws water and minerals from the earth. Within it, carbon molecules join with water molecules (CH_2O) to become a liquid. Into this special "carbon juice", it blends its own unique set of molecules which contain its DNA. As the plant grows it creates more *biomass*. Biomass is the basic stuff of life.

Through its roots, the plant releases this carbon juice into the soil, which attracts microorganisms. This is a kind of currency which plants use as barter to get the other elements they need and can't produce themselves. In exchange for some of its carbon, the plant receives nitrogen, phosphorus, potassium and other elements.

Plants are energy in solid form. Other creatures eat them in order to get the nutrients they need, which in turn helps them to grow and maintain their own biomass. So, the energy from the Sun that plants synthesize transforms it into the biomass of all living things.

Living things produce waste material, and they also inevitably die. Both their waste and their bodies are composed mainly of CH_2O. When an animal dies, its body decomposes. In this process, water evaporates, microbes and insects break down the carbon and nutrients. Some of the carbon turns back into a gas through microbial respiration, binding with oxygen and released into the air. But some of it sinks into the soil and remains trapped as organic matter. It may lie deep in the earth where there is not enough oxygen for it to turn into a gas. This is the case with wetlands and peat bogs, where decaying organic matter is trapped under water.

Layers of organic matter build up, and the weight pressure keeps it down there for a very long time. This is how fossil fuels were created. Decayed biomass was trapped under pressure deep in the ground for hundreds and thousands of years. When we harvest fossil fuels and burn them, or when we dig up a wetland or a peat bog, we expose that biomass to oxygen, causing it to volatilise and return as CO_2 to the atmosphere.

When we talk about "sequestering" carbon, we mean the amount of time carbon remains in a solid or liquid form in the earth, instead of combining with oxygen and returning to the atmosphere. Sometimes in our climate change conversations, we tend to think that we simply need to capture carbon and keep it out of the atmosphere. Technologies are being developed to store it in vaults in the earth. But carbon is part of the life cycle, taking it out of that cycle is only a partial and temporary solution. Maybe it will compensate for the speed in which we have harvested long term carbon storage such as fossil fuel, forests and bogs. But it will not restore the balance of life; only nature can do that.

Over the past hundred years, we have reduced the volume of biomass on the planet by 50%. Think of biomass as sequestered carbon; think of biomass as life. There should be more carbon stored in both living and dead things than is volatilised and returned to the atmosphere. Human activity has reduced the amount of living biomass, of which humans represent only 0.01%. According to Visual Capitalist, the entire volume of living biomass on Earth in 2020 was 1120 Gigatonnes, while the volume of anthropogenic mass, or man-made, non-organic material created between 1900 and 2020 was 1154 Gigatonnes.

We have reduced the volume of life on the planet and replaced it with stuff that will not decompose and re-enter the cycle of life. Along with sequestering carbon, we need to think about increasing the amount of life on the planet, the number of plants photosynthesising sunlight and living ecosystems storing carbon on the ground.

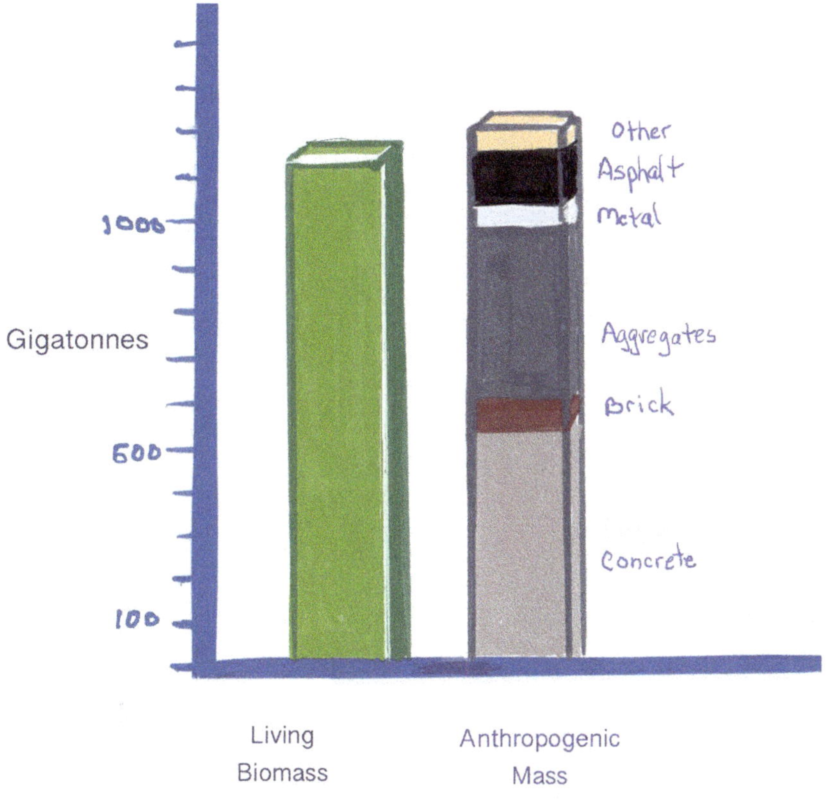

The Water Cycle

It is often said that *water is life*. Many indigenous cultures recognize the centrality of water and hold stewardship for the water as a core value in their cultural practices. Water combines with carbon to create CH_2O, the basic building block of life. All living things need it. We can survive over a month without food but only a few days without water. When living things die, water leaves their body.

Like carbon, water uses solar energy to change phase. Unlike carbon, which uses light to power photosynthesis, water requires the heat of solar radiation to convert from liquid into vapour. The warmer the temperature, the more water evaporates. As water vapour, it carries heat energy as it expands and rises. The heat is no longer sensible heat, which can be felt, but latent heat, which is stored but not expressed. The energy of the heat is carried in the water molecule. When the temperature cools, the water molecule condenses and the latent heat is released as sensible heat once again. When temperatures reach the freezing point, more heat is released as water becomes ice.

Water exists in three phases: ice, liquid and vapour. Each phase shift is mediated by temperature as well as by microbial catalysers. These processes use solar energy and are ways that nature regulates regional climates and rain cycles. Without water and its special properties, life on Earth would not be possible.

Many pioneering scientists, such as Michal Kravčík, Walter Jehne, and Anastassia Makarieva warn us that climate models are too narrowly focussed on carbon. The water cycle also plays an important role through the redistribution of heat. *Where* the heat is released makes a difference globally. They point out that the water cycle, particularly that part of the cycle governed by the evapotranspiration of plants, is the air conditioner of the Earth, and it is key to moderating temperatures on our planet.

Michal Kravčík is a Slovakian hydrologist who co-authored *The New Water Paradigm-Water for the Recovery of Climate* in 2007. Walter Jehne is an Australian climate scientist and microbiologist, founder of Regenerate

Australia. He teaches the role of the hydrological cycle to create the soil carbon sponge, sequestering water and carbon in soil and plants to mitigate the effects of climate change. Anastassia Makarieva, a Russian atmospheric physicist, is the co-developer of the biotic pump theory which is receiving increasing attention in the scientific community . This theory explains how the evapotranspiration of forests actually pulls moist air from the oceans and recirculates water through rainfall inland, underscoring how deforestation will cause vast inland areas to become deserts if we disrupt these cycles. These three scientists were some of the early voices insisting on the importance of natural ecosystems to regulate the water cycle and therefore climate. Although they each have different approaches, they all agree that mitigating the most disruptive effects of climate change must include regulating the water cycle through the restoration of natural ecosystems.

Green Water Cools

It matters where heat is transformed from sensible heat to latent heat. When evapotranspiration moves water vapour to the higher levels of the atmosphere, it takes heat with it, cooling the Earth's surface. In turn, the surface absorbs energy from the Sun and radiates it as sensible heat, warming the air further.

A law in physics, called the Stefan-Boltzmann law, states that the amount of solar energy radiated from a surface equals the temperature of the surface raised to the fourth power (Temperature x Temperature x Temperature x Temperature). This is behind the principle of thermal mass. A lot of heat is absorbed into a dark mass such as bare soil, concrete or asphalt. The surface temperature will be much higher than the temperature registered on a more reflective surface. If we use this law to calculate the amount of heat rising into the air, the amount be will much more than if the heat had first been absorbed by a plant and used to convert water vapour and then transported into the higher atmosphere, which is much colder.

Distribution of Solar Heat Energy

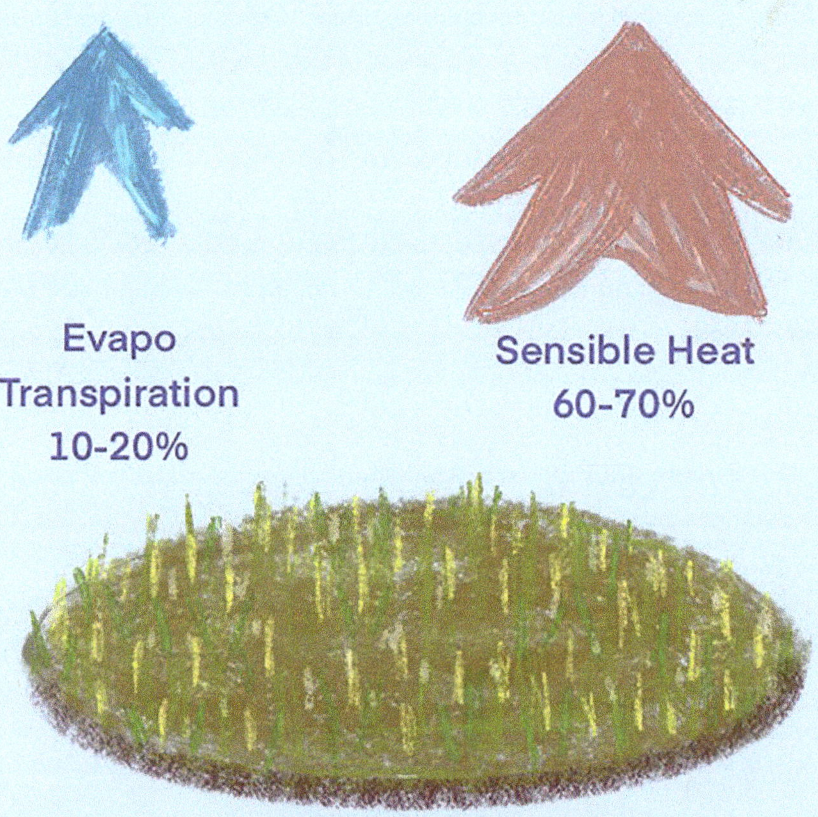

This data from Michal Kravčík of People + Water, shows how the same amount of solar energy can be distributed very differently, affecting surface temperatures depending on how much water and vegetation are in the landscape.

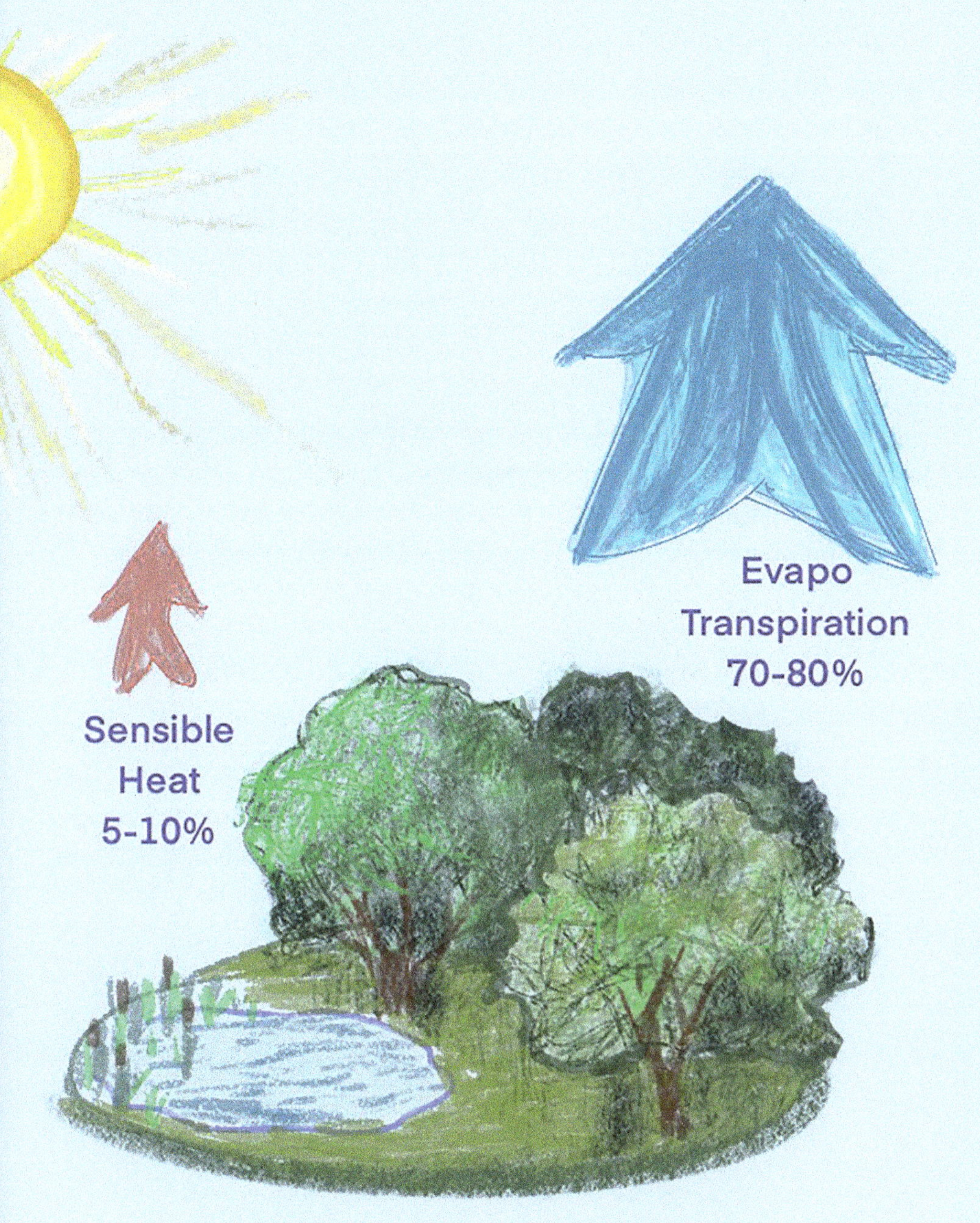

Water Retention Landscape

With water in the ground, more evapotranspiration from plants reduces surface temperatures. Much of the heat energy rises into the air through evapotranspiration.

When you fly in an airplane, the air outside the plane is very cold. The higher you fly, the colder it gets. This is because tiny particles in the air contain heat, and the closer you are to the Earth's surface, the denser these particles are. Greenhouse gasses are also contained in these particles, and there are fewer of them in the upper atmosphere because the air is thinner. This means that when latent heat is released by water vapour condensing in the upper atmosphere, the higher up it is the more chance there is some of the heat will escape into space rather than be trapped by greenhouse gasses.

Michal Kravčík calls this an energy equation. If the Sun's energy evaporates water, it will move the heat upwards, away from the surface. But if it strikes a dark mass at ground level, it will be absorbed and radiated as heat into the air. He compares this to a pot of water on the stove. As long as there is water in the pot, it boils at a constant temperature. But if you forget to turn off the burner, the boiling water eventually evaporates and your pot will get even hotter until it burns. The same is true of dehydrated landscapes; they become islands of heat. Getting water back into the soil is like turning down the burner under the pot.

Evapotranspiration enables cloud formation. Clouds form over vegetated landscapes, particularly forests where the evapotranspiration of trees transforms a great volume of water from the ground into vapour. Trees also release aerosols, such as terpenes, which cause the water vapour to coalesce into clouds.

The vapour coalesces around particles in the air, and the type of particle determines the size of the droplet. Not only terpenes, but fungal and bacterial particles also act as cloud condensation nuclei (CCN). Clouds reflect the Sun's heat away from the Earth's surface. However, if water vapour coalesces around other types of particles, such as pollutants, they form tiny micro-droplets, which create haze. Unlike clouds, hazes do not reflect sunlight back into space, rather they absorb the heat, keeping it close to the Earth's surface. So, depending on the size of water vapour particles and where they appear, they can either remove heat from the earth or trap it.

Biology Creates Rain

Clouds are required for rain, but they don't guarantee rain. First, enough moisture has to accumulate in the air to reach a critical threshold. Second, a catalyst is needed to precipitate the formation of raindrops. Dr Cindy E. Morris is a plant science researcher who has spent years studying how microbial ecology affects weather. She says that biology creates rain. Bacteria and fungi released from plant leaves or from the soil rise in the air and act as catalysts, triggering rainfall. We mistakenly think that temperature alone causes water to change phase from gas to liquid or from liquid to ice, but actually these plant-generated microbes are necessary to facilitate the change more readily. Although some inorganic molecules can also serve as catalysts, organic ones are far more efficient.

In temperate regions, ice crystals in the upper atmosphere are needed to set off rainfall. We think that the natural freezing point of water is 0 degrees Celsius (32 degrees Fahrenheit), but Dr Morris has shown that without a bacterial catalyst, water will not reliably form ice crystals until the temperature drops to -40 degrees Celsius. These tiny biological catalysts are carried through the air by winds. Dr Morris has charted the pathways that wind travels during different seasons to see where these little travellers may end up triggering rain. Depending on the pathway of the wind, there may or may not be enough catalysts to produce rain on a given day. So even in an area lying at a great distance from a forest, wind is required to bring in the organic molecules to stimulate rainfall. The more of these biological catalysts from vegetation are in the air, the more frequently and gently it rains.

A strange phenomenon was noted in south western Australia at the turn of the 20th Century. A large area of land was cleared for agriculture, but the wild rabbits so prevalent in the Australian bushland devastated the crops. So, between 1901 and 1907, the state built a 3,200-kilometre fence to keep the rabbits out. The result was a clear, the "bunny fence", as it was called, created a clear, long demarcation between the native vegetation and agricultural land. Curiously, over the agricultural land the sky is usually clear, while over native vegetation there is significant cloud cover. Consequently, the land covered by native vegetation receives more rain than the agricultural land. Nothing, aside from land use, could account for the difference in climate and scientists have been puzzled by this phenomenon for many years. There are numerous hypotheses. Dr Morris' hypothesis has yet to be definitively proven, but there is clearly a connection between native vegetation and the presence of clouds and rain, and the disruption caused by modern agricultural practices.

When it rains, some of the water flows over the ground, always downward along slopes, gathering into streams and rivers which carry it to the sea, and some of it penetrates the land. The rain hydrates the soils, nourishes the vegetation and refills the aquifers. The more rainwater that soaks into the land, the better it is for the land, animals and people.

Eventually enough water will filter into the land and hit an impenetrable surface such as bedrock. Here it begins to accumulate underground and pressure builds up. It is this pressure that forces natural springs to come up from the earth. Underground water keeps wells filled, keeps streams and wetlands from drying up in the heat of summer. Underground water keeps the plants alive and permits evapotranspiration, which moves the heat away from the Earth's surface.

Unfortunately, human activity has reduced the amount of underground water stored in most parts of the Earth. Not only does a growing population continue to dig wells and draw on water reserves, but also poor water management reduces the amount of rainfall which gets back into the earth to refill the reserves. Agricultural soils left bare and tilled with heavy machinery compact the soils, preventing deep infiltration of rainwater. Dry bare soil becomes hydrophobic, repelling rainwater and causing it to run off, carrying topsoil into waterways. Hard-packed impermeable surfaces typical of cities, such as roads and industrial areas, collect massive amounts of rainwater, diverting it into storm water infrastructures. Humanity expedites the release of freshwater into the oceans and suffers the consequences of the resulting dehydration.

Floods are a symptom of this cycle of dehydration. Excess water drains off the land quickly with sudden heavy rainfall. Compacted soils can't percolate, and human infrastructure is designed to evacuate the water as quickly as possible. Drought conditions bake the land, and wildfires cook the soil making it even less permeable, making it difficult for the land to absorb rain. Torrential rains carry away the remaining organic matter, washing whatever fertility is left into the waterways. Until we improve water infiltration across our landscapes, this vicious cycle of drought and floods will continue to deplete our water resources and drain fertility from our soils.

Biodiversity: Cascades of life

In our story of the restoration of the Loess Plateau desert, we saw that as soon as carbon and water get together, life emerges. The activity begins with microbes, gradually becoming more diverse until it forms a complex ecosystem, a cascade of life. It begins with the smallest life forms, the first to feed on carbon and water: the microorganisms.

The existence of microorganisms was discovered in the 1600's with the invention of the microscope. It wasn't until the 1800's that pathogens were discovered as the cause of disease, which led to the development of pathogen-killing antibiotics derived from moulds.

But it was only in the 1980's that scientists discovered the incredible prevalence of microorganisms everywhere. Before that, microbiologists had to grow microorganisms in a laboratory to identify and study them. With the advent of computer technology and the ability to read DNA, microbiologists were able to identify thousands of species in the environment, even if they couldn't isolate them or understand what they did. From samples of soil, water or living tissue, they analysed the DNA of the coexisting species, coming finally to astounding conclusion which is slowly revolutionising medicine, agriculture and ecology: all life is made up of complex communities of microorganisms!

When we examine the DNA of a human being, we find that only 10% of the cells in our body are human cells; the other 90% are microbial. Our bodies are colonised by microbes at the moment of birth, as we pass through the birth canal in our mother's body. The atmosphere, the food we eat adds more microbes to the community of our microbiome. Most of these microorganisms are beneficial. We need to rid ourselves of the idea that microbes are pathogens that should be killed off. A sterile environment is, in fact, the perfect environment for colonisation by pathogens, because it is unprotected. Our best protection against infection is a thriving population of beneficial microbes well established inside our body and out.

The same is true for all living things. Plants and animals all have unique microbiomes colonizing their bodies. The health of microbial communities determines the health of their hosts and their hosts' environment.

Microbes are very specialised in terms of what they do and where they live. That is why there are so many species. There are species specialised for every environment and substance on Earth. The life cycle of a microbe is short. But with the right food, right temperature, right pH balance, right humidity, they eat and reproduce at a furious rate. Their metabolism produces waste materials rapidly and releases them into their environment. These waste materials become the food for other species. But as soon as conditions change - the food runs out or the temperature varies - they stop their activity, they go dormant or die. Because the metabolites produced by a microorganism are useful to some other organism, they alter the balance of the environment, giving space to another species to grow and multiply. So, the short life of a microorganism plays a pivotal role in the succession of life.

The life of compost is a case in point. Made up of layers of waste, it ideally needs a good ratio between carbon and nitrogen. Kitchen waste is high in nitrogen, so adding a layer of dried straw or leaves rich in carbon will ensure a healthy balance. Every element you add to that compost is home to a unique microbiome, from the vegetable peels to the straw. With the warm, moist conditions in the compost pile, the microbes go to work, decomposing their favourite food. As the food in the compost pile breaks down, nutrients are released. The compost pile heats up. The next group of decomposer microbes, the thermophilic or heat loving microbes, becomes active. There are many of these, each one specialised to consume different nutrients. Conditions in the pile continue to change as other groups of microbes take over from the previous group, which goes dormant or dies.

Make a compost

There are many ways to make compost. You can buy a commercial composter or simply make one in a bin in the backyard. The most important thing is to get a balance of dried woody material and food waste or manure. There needs to be more dry stuff than wet, otherwise it will smell bad. Every time you add food scraps, cover them with some of your dry material. This could be straw, dried grass, dead leaves or wood chips. Then watch your compost evolve! It will get warm, then cool, and the bits of food will break down. Keep it covered so it doesn't dry out. If it gets too dry, add some water. It should stay moist but not sodden. Once it has cooled, you will notice worms and little insects moving in for the finishing touches.

A compost is a masterpiece of nature's closed loop cycle. From within six months to a year, your compost will be a rich brown pile of earth, which you can use to feed your houseplants or your garden.

After the thermophilic phase is over, the compost pile matures at a slower rate. The mesophilic microorganisms take over, ones who thrive in more moderate temperatures, between 20-45 degrees centigrade. In the last stages of the composting, worms, insects and fungi move in to complete the process. The final result is a rich soil amendment, full of nutrients, organic acids and molecules which stimulate plant growth. This miraculous transformation has been accomplished by hundreds of thousands of microbial species each taking a turn in one stage of the process. At each stage, the conditions created by the group of microbes at work change the environment and signal the next set of workers to take over. We call this process a *trophic cascade*.

Tropic cascades are typical in all ecological systems. Different species interact when the environment produces conditions which are favourable for them. Their interactions produce results which alter the environment and then other species come into dominance. The more an ecosystem matures, the more species diversity grows. More complex interactions between multiple species creates a richer and more resilient ecosystem. Remove any one species and another will take over the job. But if you remove too many, leaving gaps in the ecosystem, things will break down.

Imagine moving into a new neighbourhood. At first, you don't know anyone. Life is a bit lonely and difficult. But then you make friends and it gets easier. The longer you live there, the more connections you make. You know the best places to go, you have many friends and acquaintances. Life becomes easier, richer. A good community with many social connections makes for a good quality of life.

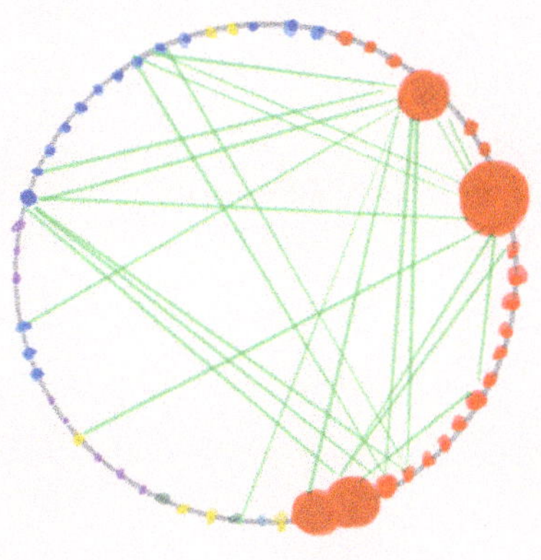

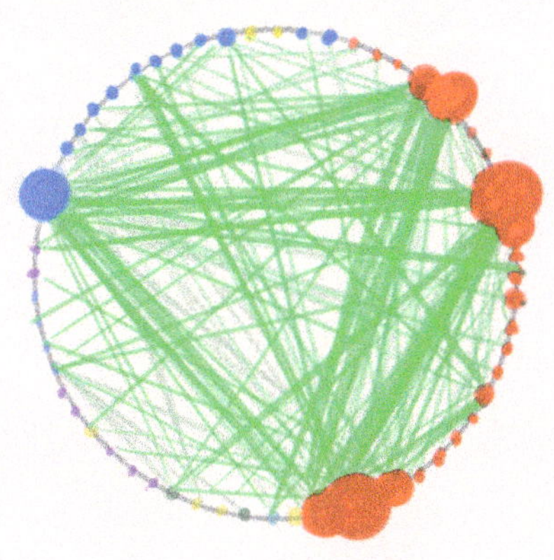

Recent ecosystem   Mature ecosystem

Ecosystem Connectivity :
The dots along the outside circle represent different species in the ecosystem. Different colours represent different families. Larger dots represent larger populations. The green lines are interactions between species. We can see that as an ecosystem matures, its connectivity and richness increase.

String exercice

Have the group stand in a circle. Give one person a ball of yarn and ask them to hold one end of the yarn and pass the ball to another person across the circle. Have the participants pass the yarn randomly back and forth around the circle while continuing to hold their bit of yarn. After a while the whole group will be connected by an intricate web.

Explain that every connected strand is a relationship. Now the facilitator can take a pair of scissors and cut one strand at a time. The more complex the web, the longer it takes to break apart. With only a few connections, the web would be broken with only one or two cuts.

Chapter Two
Soils, Grasslands, Forests & Wetlands

Nature has evolved systems to sustain the three pillars of life everywhere on the planet. Each of these types of ecosystems has evolved to be suited to the climate of a particular part of the Earth and each serves the carbon cycle, the water cycle and biodiversity in their own way.

The development of the modern, industrial society has provided humankind with huge benefits but has also removed a huge swath of natural ecosystems. We are now feeling the consequences. For millennia, they have provided us with fertile soil to grow food, clean potable water and breathable air. Although going backwards is not an alternative, we can learn how healthy ecosystems function, and use them as models in adapting our practices. This is what agroecology, Nature-based Solutions and biomimicry are all about.

The soil ecosystems

Soils are the foundation of life for terrestrial ecosystems, and we are losing them at an alarming rate. Soil is not just the inert mineral component of rocks reduced to tiny particles. Those account for less than 50% of healthy, living soil. The other half is air, water and billions of microbial organisms. If you pick up a handful of healthy soil, you will be holding more living beings than there are humans on Earth. The soil will be teeming with bacteria, fungi, tiny decomposer insects and worms, all microscopic.

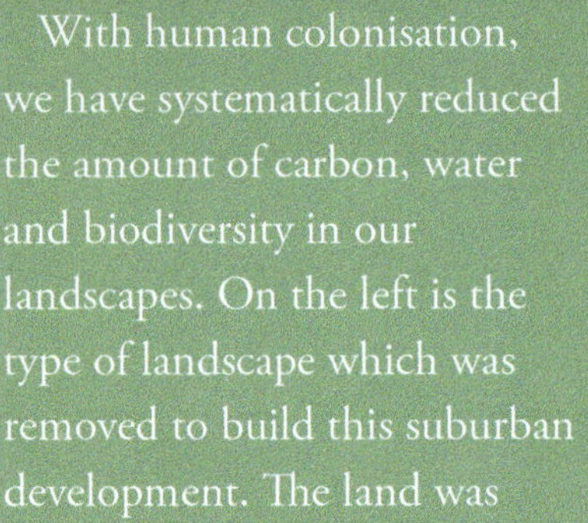

With human colonisation, we have systematically reduced the amount of carbon, water and biodiversity in our landscapes. On the left is the type of landscape which was removed to build this suburban development. The land was cleared and the biodiverse living system was replaced with inorganic matter, with the exception of a little patch of lawn. How much carbon, water and biodiversity were displaced and what value do we give to those pillars of life?

Soils require plants, because without them, they will not be healthy for long. It will dry out and lose its structure. The air, water and the living microbiome cannot be sustained without their input. Plants and soil live in synergy with one another; plants feed the soil and soil feeds the plants.

Plants use the Sun's energy to transfer liquid carbon through their roots to the microbes in the soil in exchange for nutrients. But there is more to it than that. Plants and microbes use a secret chemical language to communicate. Every species has a unique microbiome, and these biomes produce phytochemicals. When a plant needs nitrogen, for instance, it produces a chemical that is added to the carbon juice coming from its roots. That chemical is a signal understood by particular microbes which convert nitrogen from the air and share it with the plant. Each species selects and feeds its own particular cohort of microorganisms. Some microorganisms are more generalist and can be shared by many plant species.

Because each has its own particular microbial helpers, the number and variety of plants determine the population of microbes in the soil. It also works the other way around. For example, trees prefer fungal helpers, so in the forest and beneath trees, the soil is dominated by fungus that likes to associate with tree roots. If you plant a sapling in the middle of a field where there are few trees, it is unlikely to survive unless you add soil from an environment where that species of tree grows, a soil rich with its particular microbial helpers.

The right balance of bacteria, fungi, microscopic insects and worms is needed for healthy soil, but their ratios depend on the type of plants growing in the soil. The microbial communities in modern day soils are often unbalanced because modern agricultural practices such as tilling and the use of agricultural chemicals are very hard on many microbial players, especially fungi and insects.

Can you imagine your city being turned upside down, shaken, and then bombarded with poison? That is what it is like for a microbial community to be tilled and then sprayed with chemicals. Most of our modern-day agricultural practices were developed long before the discovery of the microbiome. When industrial agriculture was developed after World War II, soil was thought of primarily as a medium for stabilising plant roots.

The nutrients were delivered in the form of fertilisers. Insects were considered to be pathogens, so insecticides were developed to kill them. Unfortunately, when plants get their nutrients directly from inorganic fertilisers, they don't send out chemical signals to the soil, to tell the microbes they need nutrients. The microbes which could do the job for free aren't fed. If fertiliser stops being added, the beneficial population of plant helpers is no longer available to carry out the task. It takes time to rebuild the healthy populations in the soil to restart the closed loop cycle.

Industrial agriculture systems gradually degrade microbial communities in the soil. The plants grown in these monocultures require more and more chemicals to fend off disease. More diverse plants with their helper microbial communities make for more nutritious food and stronger soils. Providing habitat for pollinators and other beneficial insects reduces the need for pesticides because nature finds a balance. When there is a strong balanced community of beneficial microorganisms, the disease-causing pathogens are few in number. The less diverse a system, the more fragile it is.

Why is soil important?

Aside from the obvious need for soil to grow our food, it is also critical to the healthy balance of the carbon and the water cycles. As a natural system, soil is nature's recycler, the original closed loop system.

Soil that is well fed with organic amendments is generous in productivity. In modern conventional agriculture we harvest biomass (crops grown in the soil) without giving anything back. Although fertilisers and other chemicals help plants to grow, they bypass the natural fertility cycle.

Over time, these soils become depleted and require higher doses of chemical inputs to maintain production, like addicts who need higher doses of their drug to get the same effect. A soil devoid of its balanced microbiology lacks structure, compacts easily and fares badly in extreme weather conditions. Our agricultural soils are gradually being depleted of their carbon. While soil has the potential to store up to 10% organic matter, modern agricultural soils average less than 2%.

But we can feed soil to increase its microbial diversity and biomass. What does soil eat? Plant debris, preferably a diet of diverse plants, plus animal waste. On regenerative farms, farmers plant crops specifically to feed the soil. They don't remove everything without giving back. Compost, manure or compost extracts all serve this purpose. They may plant "cover crops", deliberately to protect the soil, keep it moist and add nutrients.

When microbes in the soil benefit from a diverse diet, it is rich in "soil organic matter" (SOM).

Carbon is stored along with a healthy range of bacteria, fungi, and microscopic decomposers, maintaining its structure, which ensures that water and air are stored in soil particles. With every percentage point increase of SOM in soil, water holding capacity increases dramatically. Drive by the average farm after a heavy rainfall, and you will likely see puddles in the fields. You might also see low-lying areas where the crops are stunted and yellow because the water is not draining properly. But water on a farm with healthy living soil will sink in deeply and quickly. It will refill the water table and be stored to help the plants survive dry periods.

Healthy living soils store carbon and water. They are critical for our food security in a changing climate.

Vivian Kaloxilos: Ecologist and Soil Doctor

Vivian Kaloxilos runs a business in Quebec, Canada, called Docterre. Her passion is to help people understand and optimise the soil ecosystem. Vivian is a trained ecologist and believes that the ecosystem of the soil is the foundation of all other ecosystems. Most people have little appreciation for its importance, even farmers and horticulturalists who earn their living from it.

Vivian's journey to becoming a soil doctor began at university. Her anxiety over the environment prompted her to throw herself into activist causes. One day, during a student demonstration, she had a headache. A fellow activist offered her some peppermint oil to alleviate it. Somehow the notion that one could heal oneself with plants was a lightbulb moment for her. As plants are part of the environment, she decided to go into an environmental program at the university. The program gave her a broad view of ecosystems and their function. Yet it was very academic and focused on large environmental issues. The approach left her feeling powerless sand depressed. Fortunately, through university connections she met people with more solution-oriented and inspiring ideas, many of whom were students of permaculture.

Permaculture is based on the principles of natural ecosystems. The word means "permanent culture", a self-sustaining system, like nature itself. The design concepts of permaculture can be applied to anything from agriculture to architecture to wastewater. Inspired by these ideas, Vivian pursued her own explorations into herbalism, ecological agriculture and applied ecology. She completed her degree and decided to become a market gardener.

Her first requirement was land, but she lacked enough capital to buy any. Through a contact made in her network of activists, she was invited to join an eco-village project. The group had just acquired an eighty-acre farm, which they named Valhalla. It had been a conventional farm for forty years, growing GMO maize and soybeans. As a result, the soil was severely degraded. Before anything could be grown without chemicals, the land needed to be regenerated. Vivian compared the soil to broken pottery, and said the ecosystem was nothing more than a horde of mosquitos. She felt called upon to heal the land. At first, she was just experimenting, putting into practice what she had learned or read. Although she lacked experience, she grew cover crops, made biodynamic brews from plants, and imported organic matter.

Gradually, using her basic knowledge and intuition, she learned to listen to what the land was telling her to do.

Dr Elaine Ingham is an American biologist and founder of the Soil Food Web, after a term she coined for the many creatures involved in cycling of organic matter in the soil. Vivian met Dr Ingham at a permaculture event in 2013. In their discussions, Vivian realised that the methods she was employing at Valhalla were actually about nourishing the microbes in the soil. Vivian began to study with Dr Ingham and soon became one of her protégés.

The Soil Food Web teaches students the unique role of each player in the Food Web. Students use microscopes to verify whether a sample of soil or compost is active and balanced. Most importantly they learn how to make good compost and what is known as "compost tea". Dr Ingham is the guru

Left: Vivian makes high quality, balanced compost and uses
a compost extractor to make a biologically rich inoculant.

of aerated compost tea, and she swears it is the cure to just about everything. Aerated compost tea is made by taking a small quantity of high-quality compost, brewing it in water with added nutrients to feed the microbes, after which the mixture is aerated with a bubbler for 12 to 24 hours. The microbes in the compost rapidly reproduce, making enough compost tea to spray over a large area. This method is more efficient than spreading compost over a field.

Vivian applied her Soil Food Web training to her work at the Valhalla farm. Within three years, the farm was teeming with frogs, snakes and birds. Every year biodiversity on the farm increased. The soil improved. Vivian no longer aspired to start a market garden, although the soil was now healthy enough for her to do so. Few people, including farmers and ecologists, understood the microbial basis of soil life. She was inspired to share her knowledge and skills but wasn't sure how. While she continued the work at Valhalla, sharpening her skills, she performed as street musician to earn money. At the same time, she offered her services as a consultant, giving guidance on soil health and compost, but without much success.

In 2015, despite her initial reluctance to become an entrepreneur, she started her own consulting service, Docterre. She bought a microscope and began offering workshops to community and permaculture groups. Trainers hired her to teach the modules dealing with soil in their courses. She advised private clients on making compost and applying soil regeneration methods in their gardens, which eventually led to her being awarded a large contract to treat seventy residential lawns in Ottawa. She brought in her specialty compost made at Valhalla and hired an assistant to help her make it into compost tea. Each lawn received several treatments over the course of the season and the results were good. The treated lawns fared better than their neighbours' through the dry, hot season. But treating lawns was not what Vivian wanted to do ultimately. She wanted to get into agriculture.

Vivian got a break in 2017 that brought her greater recognition. She was one of a group of five women who collaborated to organise a major event, The Living Soils Symposium, in Montreal. Together they managed to raise enough money in grants and sponsorships to attract some well-known international speakers and an audience of over 400. The theme was how soil management could help in the climate crisis. Speakers included Regeneration International, who argued that regenerative agriculture is necessary to restore climate stability, Project Drawdown , which had just published a book ranking 100 climate solutions, and the French-based international "4 per 1000" initiative , whose representative pointed out that increasing the amount of organic matter to 4 parts per 1000 (0,4%) in the world's arable land could draw down all the surplus carbon in the atmosphere.

The success of the Symposium gave Vivian the visibility she needed. She began working with large-scale farmers and with "early adopters" keen to improve their soil without the aid of chemicals. She helped the owner of an orchard move away from spraying her trees with sulphur to spraying them with compost tea. She taught farmers to make their own compost and encouraged them to abandon tilling to control weeds.

Soil undergoes a succession of stages in ecology. Depleted, bare soil tries to cover itself with annual, early successional weeds. If the successional process continues, the soil will become de-compacted and start to build up organic matter. The next plants that grow in this soil will be woody, perennial plants. The soil will progressively develop a higher ratio of fungus. Left alone for long enough, it might turn into forest, with soil primarily dominated by fungus. But by tilling, modern agriculture confines the soil to its early successional phase, with weeds constantly struggling to take over and de-compact the soil. Vivian found it difficult to persuade farmers to adopt the techniques she was proposing, as they lay outside the framework of modern agronomy.

So, in 2019, she resumed her educational training courses for farmers. Vivian feels that if farmers are to lead an ecological transition, they need to be eco-literate, taking charge of their farms rather than relying on consultants. They need to go beyond an agronomic understanding of plant nutrition to a deeper understanding of soil ecology. It was also clear that few farmers had the time to make compost, and very little good quality compost could be bought on the market which would produce the exceptional results that Vivian had achieved. Today, she sells her compost to farmers to make compost extract, an efficient and simplified alternative to compost tea, without the long aeration process. A bucket or so of good compost can be turned into a thousand litres of spray, enough to treat eight acres of land.

While compost is an important tool, it is not the only one in the toolbox. Soil microbes need plants, and plants need soil microbes. A greater diversity of plants calls forth a greater diversity of soil microbes. Vivian now teaches functional ecology for the entire farm ecosystem.

Each plant species has a different function in an ecosystem. Some fix nitrogen, for example, while others de-compact the soil. Furthermore, some are more drought resistant than others. As farmers select plant species to include in the rotation process or in a cover crop mix, they look at increasing *functional* biodiversity. If we prepare for every contingency, the whole system will be more efficient and resilient. The more different the plants are from one another, the more likely they will work in a complementary way.

Once the soil is restored with a healthy web of fungi, each plant shares its unique chemistry through the underground connections of the soil food web. If there is a drought resistant plant in the mix, it shares its genetic expression with the rest of the ecosystem. When farmers add fertilisers to the soil, even organic ones, plants absorb the nutrients but invest less in forming relationships through the soil to get the nutrients they need. The plants become lazy and the soil ecosystem suffers. If a soil has a good layer of hummus, enough organic matter and functional plant diversity, plants should be able to extract by themselves all the nutrients they need.

Vivian has built a human ecosystem of regenerative farmers. They apply her methods, and purchase high-quality compost, or produce it themselves. To reach more people, together with her partners she puts training courses online. By accompanying this new cohort of regenerative farmers, she has built up a store of inspiring stories that she can share with others. Soon perhaps, she will be well enough established with Docterre that she has time to get back to making some music on the side.

Félix Noblia: Regenerative farmer and influencer

Félix Noblia has been on a long learning journey since he bought his uncle's 150-hectare farm in the south of France. He knew nothing about farming at the time. When he took over in 2008, it was a conventional farm with cows and silage corn, planted using deep tillage and the full spectrum of fertilisers and pesticides. Cattle were confined to a single pasture for six months of the year. The soil was depleted and the farm was unprofitable. Félix's motivation to change methods was at first economic. Payments had to be made, and the cost of continuing to farm in the conventional manner was prohibitive. As he had no background or training in agriculture, he came to it with a freshness and a desire to learn and experiment. Together with agronomist Emmanuelle Bonus, they set out to find different ways of farming that would make the farm profitable again.

The first step was to repair leaks in the system. The soil was losing nutrients from inefficient management. He decided to diversify his crops. Maize, because of the high cost of inputs, was not a solid investment. Instead, he started planting a greater selection of crops and shortened the supply chain with direct sales. He grew wheat and sold it on the local market, and planted sunflowers and canola, which were purchased by a regional company to press for oil. Félix and Emanuelle Bonus continued to experiment with diversifying their crops further. They were influenced by the American movement of "conservation agriculture", a method developed to minimise soil erosion and increase its carbon content.

In 2011, Félix abandoned ploughing and experimented with cover crops. In 2013 he transitioned to no-till farming, a process of seeding the crop directly in the debris of the previous crop. The debris covers the soil surface, reducing the amount of weeds and keeping moisture. Gradually it decomposes, adding

organic matter which feeds the soil. With no-till farming, the soil is never turned or left bare. In conservation agriculture a previous crop is either killed off by winter cold or herbicide. Special planters are used to drill the seeds through the plant residue and into the soil. The challenge for no-till farmers who don't use herbicides is combatting perennial weeds. Félix is one of a growing number of pioneering farmers who are refining their methods to make this method viable without recourse to herbicides.

In 2015, he adopted a holistic method of managed grazing for his cows, installing movable electric fencing to control their movements. The herd is confined to a small area for a short time, being moved frequently to avoid overgrazing and to distribute their manure and urine over the field. Although we often hear about the environmental damage caused by overgrazing, using ruminant animals in this way is actually one of the most efficient ways to build soil. In a short time, Félix doubled the productivity of his pastures using this approach. Healthy soil meant that grass was growing quickly and densely enough to provide twice the amount of feed for his cows. He also planted rows of mulberry and nut trees along the fields to feed the soil

network as well as provide his cows with shade. If he wasn't so busy, he could harvest the fruit from these trees for extra revenue. The main reason for planting them was to increase the diversity of the farm ecosystem. The combination of trees and grazing animals is one of the most effective ways to increase the water and carbon stored in the soil.

The following year, Félix decided to apply for organic certification. This was a big challenge, because no-till farmers usually rely on herbicides to terminate their crops and control invasive weeds, while organic farmers rely on tilling. Avoiding both herbicides and tilling is a challenge that only a small number of the most innovative farmers have dared to tackle. Instead of herbicides, organic farmers use a roller crimper: a huge heavy roller with ridges that crush and kill the cover crops in order to plant the next crop. Many organic farmers, including Félix, will adapt existing equipment for their own purposes. Felix's roller is filled with water and applies 1000 kilos of weight per metre. The permanent ground cover of crop residue keeps most weeds under control, although a shallow till may need to be carried out every few years.

Using no-till and cover crops, especially in organic farming, is an exercise in complexity, not only for the management of weeds. Nitrogen-fixing crops such as clover and beans must be alternated with cash crops that are hungry for nitrogen. This is more complicated than just applying a prescribed amount of nitrogen fertiliser at the right time. Nitrogen-fixing crops release it slowly over a long period, so it takes time for the system to become efficient. The possibilities are endless, depending on the region. Deciding on which plants to alternate with cash crops and what season to do so means that every farm is a laboratory.

Small pockets of passionate and courageous farmers are doing this work all over the world, trying to build up enough experience through trial and error so that they can help other farmers make the transition with less risk. Félix is among the most passionate and daring of that group. He has managed to make it work and reduced the cost of inputs on his farm by 200 euros per hectare.

In recognition of his pioneering work, he was awarded a trophy from the French Minister of Agriculture for innovation in agroecology in 2016-2017. The award brought him a lot of media attention, with the result that the Department of Agriculture, which at the time feared a change of government less open to these innovations, asked Félix to spearhead a public awareness initiative on the benefits of agroecology. Félix was young, attractive, well-spoken and successful in agroecology. The government trained him to become a social media influencer. He now has a Facebook page, YouTube channel, has presented a TED Talk, and became Vice President of Fermes d'avenir, an association promoting agroecology.

Over the years, Félix noted a decrease in rainfall and fewer insects on his farm; the climate was becoming more arid. He began to feel very concerned about climate change. He started planting sorghum in his fields. This dense drought-resistant grass would ensure that his cows could graze in the hot

Félix Noblia' roller crimper, a heavy roller which crushes the cover crop so that the next crop can be planted through the plant debris.

summer months. All this invoked a sense of urgency in him that larger scale change was needed rapidly. He considered that his own regenerative work was part of the climate solution, but unless the work was carried out on a landscape scale, his efforts would make little difference.

This sense of urgency motivated him to set up Régénération. The goal of this non-profit company is to create an ecosystem of farmers, agronomists, businesses and investors to finance and support the transition of 4-5 million hectares of agricultural land to regenerative farming within six to seven years. They developed an investment product based on the ecosystem of healthy land (carbon, water and biodiversity). In addition, they deployed a method of evaluating and verifying these impacts as well as a network of skilled consultants to train and accompany farmers, partly through financial incentives, to make the transition. Félix believes that a massive transition on a landscape scale to regenerative farming will mitigate climate impacts and provide food security.

Grassland ecosystems

There are some parts of the world which are naturally arid, areas which historically get low rainfall, either because of their high altitude or because they are far inland away from moist sea breezes. It could also be on the inland side of a mountain range from where the rain falls. These parts of the world do not get enough rainfall to support the growth of forests. So, nature has evolved ecosystems which can still cycle carbon, store whatever water there is and sustain life.

Grasslands and savannas support herbaceous plants such as wildflowers, woody shrubs and various grasses. The creatures that feed on these plants are mostly ruminant herbivores: goats, cows, sheep, horses, deer, buffalo, camel, zebra; birds and insects which live in symbiotic relationship with these plants and animals. Many birds, for example, follow ruminant animals to feed on the insects that spawn in their manure or live in the animals' hides. And of course, there are the predators to keep the ruminant population in balance.

Humans have destroyed many of these grassland ecosystems and consequently the diversity of species which inhabited them, mostly through annual farming, with the resultant loss of soil carbon, water reserves, and underground aquifers. Annual cropping systems are typically low in diversity, tend to compact the soil through the use of heavy machinery and frequently require irrigation. As farmers convert the land to annual cropping, which lacks resilience to extreme weather, these arid regions become increasingly vulnerable.

Historically, grasslands were home to herds of grazing animals, which moved across the landscape, eating the vegetation, stimulating its growth, fertilising the soil with their manure and urine, and trampling the ground, helping the dry thatch to decompose and return to the earth. In most traditional pastoral cultures, ruminants roamed freely, attended by nomadic shepherds. Problems arose when humans confined herds to one area. They overgrazed, eating the grasses down to their roots, eventually destroying them. Overgrazing has been the cause of desertification of many regions, because humans have disregarded the animals' important role in sustaining the ecosystem.

In the regenerative movement, innovative ranchers try to reverse the degradation by managing grazing animals in accordance with their natural behaviour.

Ross Mac Donald: Champion of grassland ecosystems

Ross MacDonald lives in southern Saskatchewan, Canada, one of these natural grassland regions. Saskatchewan is situated at the northern tip of the central plains area on the dry side of the Rocky Mountains range, which lies at the centre of the North American continent. Ross is a pioneering voice for grassland ecosystems.

He grew up on a farm outside a small town in Saskatchewan. His parents were lawyers, not farmers, although they owned a few hundred acres on which they grew annual crops, hay, and raised some cattle. His maternal grandparents had been raised on homestead farms in Saskatchewan in the 1930s.

Because they lived close to the land, they were aware of the hardships farmers faced. During the 1980's, farming in Saskatchewan went from being a lucrative concern, thanks to adequate rainfall and high commodity prices, to a risky one as droughts set in, prices plunged and interest rates rose. Many farmers sunk into debt, and much of Ross's parents' law practice was devoted to helping farmers save their business and keep their land. Ross's father noted the loss of native grasses during the 1970's, when government programmes and high commodity prices encouraged farmers to plough the native prairie to grow wheat and other annual crops on marginal soils, ill-suited for such use. Within ten years, the soil eroded away, leaving little to hold the scarce water. Ross's father would bring him along as a boy when he visited clients and discussed the importance of sustainability and the environment in managing a farm. Ross didn't realise until much later how formative these conversations had been.

As Ross got older, he had the opportunity to work on the ranches of some of his parents' clients. He felt in his element among the animals, out on the land. He decided to pursue a career in animal science, and enrolled at an agricultural college in Saskatoon. As graduation approached, he saw that his job prospects were limited to working in a pig barn or selling veterinary pharmaceuticals, neither of which appealed to him. After a fellow student gave him a journal of range management, full of photographs of people riding horses on the open land, carrying out research in animal behaviour, he decided to apply to Montana State University for his graduate work.

Montana, which lies just over the US border with Saskatchewan, has a strong ranching tradition. Ross lived and worked on a ranch associated with the university while he completed his Masters project on dominance behaviour in cows. The blend of academic research alongside the day-to-day work of ranching was ideal to him. It became the vision of what he wanted to do upon graduation.

There were few opportunities in Saskatchewan for this type of work, so Ross took a job with a small organisation dedicated to wetland conservation. Part of the organisation's funding went to grassland and habitat restoration for prairie species at risk. His work involved advising farmers and ranchers on the importance of grasslands, building up of the soil and of wildlife habitat, and on creating "good news" success stories. The job fulfilled the academic side of his vision; he just needed his own patch of grassland to complete it.

There was a lot of land available for sale in one of the most arid parts of southern Saskatchewan, an area with sandy, gravelly loam soil. Unfortunately, Ross lacked the capital to buy it. But he had experience with conservation organisations. One of the ranches in Montana where Ross worked had used conservation easements to expand the number of acres they managed. These are legal agreements between a landowner and a government agency or land trust that sets limits on land use in order to conserve the environment. At the time, these were new in Canada, and the legal framework was still being worked out.

Following discussions with the Nature Conservancy of Canada, Ross became one of the first ranchers to adopt a conservation easement. The value of the easement would finance his downpayment for the ranch, which he called "98 Ranch". He bought 280 acres of land. Today, his total acreage grazed is close to 5,000. It is a mixture of deeded and leased lands, 90% of which are native prairie grasslands. Ross has formed partnerships with a number of conservation groups, which have helped his business, expanded his knowledge and established personal relationships. Since conservation easements are contracts to preserve the ecological integrity of an area into perpetuity, Ross agreed to preserve the grasslands and not to drain the wetlands, choices completely in line with his values.

When Ross bought the ranch in 2001, he had only a dog and a few horses. He struck a deal with his neighbour, whose cows would be allowed to graze on Ross's land for the season, which would ensure both the health of the grassland and a long-term revenue. The land was divided into eight sections,

five of which had native grass. Two sections — a total of three hundred and twenty acres — had been used to cultivate annual crops, and one section, previously used for annual crops, had been reseeded with perennial grass, which was harvested for hay. The damage caused by tilling had resulted in a shocking loss of topsoil, particularly on the hilltops. Although it is possible to restore native grasses, the process is long. The diversity of species in those pastures is incomparable. From tiny sedges to the much larger species, they all provide necessary ground cover. They are never bare, even where the soil is thin and gravelly, or where it has been heavily grazed. These spots are covered with mosses, lichens, forbs (broadleaf, flowering plants that grow in meadow ecosystems) and other pioneer species which protect the remaining soil from washing away. In replanted areas, that biology has been destroyed, so while perennial hay species can be grazed, they are not nearly as robust and resilient as native ones. After 20 years of careful management, the natural biology has begun to return in the reseeded areas.

The art and science of grazing management for grassland restoration involves "reading the disturbance". For plants, like humans, a certain amount of stress is a good thing. In the case of grasses, it stimulates growth. Too much disturbance, however, slows growth, and the grasses need rest to recover. The more disturbance, the longer the recovery time. From observation and experience, the careful rancher knows when a field needs rest from grazing. With the right balance of disturbance and rest, the land will store more carbon and water, and support more cows, more frequently.

Ross's conservation job outside the ranch allows him to leave his reseeded fields and native prairie pastures to rest for a year or more and re-establish their natural biology without the pressure from grazing. He respects the successive stages grasslands go through on their recovery. For example, he allows a severely eroded field on a hill, which has been reseeded, to sprout native forbs, instead of planting a non-indigenous but more desirable grass for the cows. Ross believes that if you co-operate with natural processes, you are no longer managing the land, you are learning to let the land manage you. He observed that fields re-seeded with non-native grasses became infested with gophers and badgers, which disturbs the grass. Ross suspects that the animals are attracted to compacted soil because it makes for stable tunnels. Another rancher might try to wage war on these burrowing mammals, but Ross noticed that after they had moved on, their digging had de-compacted the subsoil, stimulating the regrowth of early successional native plants. He believes that these creatures occupy an important ecological niche not yet fully understood. For this reason, he doesn't fight gophers.

As we saw in Félix Noblia's story of holistic managed grazing, moveable electric fencing is often used to confine a herd to a relatively small area in order to control their movement across the land. Ross uses this technique to some extent, but he also works larger pastures with traditional herding techniques using horses and a dog. What is important is the result, which is achieved by refining one's ability to read the landscape and to adapt. Tensions often arise between traditional ranchers and newcomer proponents of holistic managed grazing, but Ross wants no part of it. Methods are just tools. He is on good terms with both groups. His aim is to restore the land and make a living doing what he loves. He has managed to stabilise the grasslands on his ranch so that they fare pretty well through the frequent dry summers and the occasional flood.

Until recently, the marketplace did not always reward a rancher like Ross for his efforts. However, a movement is growing, led by a few big companies, to gain recognition for the improved quality of meat raised in a natural environment. In 2012, Ross responded to ads in certain trade journals that were seeking producers who minimised the use of hormones and antibiotics. Ross's ranch isn't organic, but because the animals graze on a natural grass-based diet, and their nutritional demands are co-ordinated with the grassland's cycles, the use of hormones and antibiotics is rarely necessary.

Ross was offered a premium price for his high-quality meat, securing a contract with A&W. The fast-food chain had begun a marketing campaign to source natural beef, since consumer preference for increased options had expanded market demand.

Ross was featured in several of the company's promotional videos and developed a relationship with staff members which turned out to be mutually rewarding. He gained a market which rewarded the added value of his work, plus an education on corporate marketing, digital production processes, supply chain dynamics and end product specifications. And A&W learned a lot from a grassland expert. This relationship spawned a network of other cattle producers interested in this type of production. At first, the focus was exclusively on *no hormones, no antibiotics*. But when the film crews came to his ranch, Ross talked about the grass, insisting that the quality of his meat was due to the grass, much to the annoyance of the ad team, who wanted him to focus on the management of the cattle. On the next visit, he again talked about the grass, as well as the signs of a renewed healthy ecosystem, such as the appearance of songbirds. The crew was a little more receptive. By the time of their third visit, they had modified the focus of the campaign from *no hormones, no antibiotics* to *the benefits of grassfed beef.* When the film crew came to shoot the video of Ross and his wife, Christine, riding horseback with the cows through the hilly grasslands, the crew admitted that they had changed their ideas. They finally understood that the health of the animals and the beauty of the landscape depended largely on the health of the grassland ecosystem where they grazed. *"Grass fed"* **actually means that the meat is not only more nutritious, but that the animal has spent its life improving the ecosystem and building climate resilience.**

Left: Native prairie grasses with their huge diversity of species thrive in this undulating landscape. The soil is protected from erosion and evaporative loss, enabling biology and moisture capture.

Forest ecosystems

The world has lost another 10% of its total forest area in the last 30 years, according to the 2022 FAO report *The State of the World's Forests*. They are being razed to make room for agriculture or development. Before humans came to dominate the planet, forest covered 57% of the Earth's terrestrial surface. Today, one-third is still forested, but we will destroy what remains if current trends continue. Climate stresses, such as warming beyond the adaptive capacity of species, insect and disease migration, wildfires due to drought, are of major concern, yet deforestation is mostly caused by human activity.

What value do forest ecosystems provide? If you think about biomass as synonymous with life, the density of life in a natural forest is greater than in most other ecosystems. Every tree, which is the tallest of all plants, has an equal amount of underground biomass in its roots as it does above ground. The rhizosphere around the roots hosts symbiotic creatures; between the trees, a myriad of plants and animals thrive at all levels, from beneath the forest floor to the canopy. Forests are home to most of the world's terrestrial biodiversity. All this life is a storehouse of carbon and water. Every bit of forest lost represents the vaporisation of the carbon and water housed in those lives.

Aside from the living biomass that a forest represents, it is a huge driver of the water cycle. It preserves water in the bioregion (an area defined by natural rather than human boundaries) by cycling it through evapotranspiration and the rain cycle. A single mature tree can cycle more than 200 litres of water in a day, removing heat energy from the surface air. Multiply that by the number of trees in a forest and you are talking about a massive volume of energy moving a huge volume of water into the atmosphere. Forests create clouds and rain. It is common to see clouds of vapour above forests, which may be retained under the canopy to hydrate other plants or escape it and later fall as rain. Or it may be blown away by wind to water an area down wind. Forests protect the ground from drying out because they contain the water and allow it to penetrate the ground. The roots and their companion fungi make a deep, moist, carbon sponge.

Suzanne Simard: explorer of forest ecologies

Suzanne Simard is a professor of forest ecology at the University of British Columbia in Canada. She has gained international recognition for her ability to communicate the discoveries she has made in explorations of the intimate life of the forest.

Suzanne comes from a family of several generations of loggers. She was born and bred loving trees. She spent her childhood exploring the forest and couldn't imagine living away from it. Forestry was a natural career choice. But as a young woman in a male dominated, increasingly industrialised sector, she had a difficult time. Like any corporation, this industry is governed by the drive to maximise profits. It is common practice to "clearcut" large swaths of forest, which means cutting down most of the trees and replanting monocultures of a quick growing, commercially desirable species. The idea behind clearcutting is that competition from other plants for light and resources will compromise the growth of the cash crop. Therefore, the liberal use of herbicide in the tree plantations is standard practice. The policy is called "Free to Grow" meaning that any plants other than the desired species are considered unnecessary competition.

Suzanne felt uncomfortable with these unnatural forests, but never voiced her discomfort among colleagues, afraid that it might isolate her. She noticed that many of the fir seedlings she planted were yellow and did not thrive in the plantations. Because of her childhood experiences examining the masses of fungal networks she would dig up around tree roots, she suspected that the seedlings missed the presence of their fungal allies. Yet the industry was resistant to any suggestions of change, especially from a newcomer who also happened to be a young woman.

After struggling to find her place for several years, she left the logging industry to go into the British Columbia Forestry Service Research Department. There, her experiments on fir trees demonstrated that using various amounts of herbicide to kill off competing vegetation had no long-term benefit sind was, in fact, detrimental to the fir seedlings. She went

Next double page: Large trees support a diverse ecosystem both above and below ground. This nurtures and protects the growth of new younger trees and their communities of helpers.

on to design a series of experiments to study the relationships between pines and firs, species desirable to the forest industry, and native plants growing alongside them. Using chemical markers, she traced the exchanges of nutrients between the plants and discovered that, not only did the fungal network help plants to grow, but it transported nutrients and biochemical messages between them.

Furthermore, she found that certain species which commonly grow together have symbiotic relationships. Nitrogen-fixing alders transfer it to pines. In plots where alders had been completely eradicated, the weakened pines succumbed to disease and insect infestations. The species of mycorrhizal fungus that colonise the roots of nitrogen-fixing alders also connects to the pine roots, enabling the direct transfer of essential nitrogen.

She then looked at the relationship between fir and birch. Since birches compete for light in fir plantations, they were eliminated with herbicide. Yet Suzanne noticed that in the cleared patches, many of the firs were dying from Armillaria root disease. She discovered that there was a flow of carbon between fir and birch. In the summer, when the birches were at their peak of photosynthesis, they shared as much as 10% of their carbon with the fir. But in spring and fall, when the loss of their leaves prevented the birches from photosynthesising, the firs sent carbon to them. There exists a complex web of communications that allows tree species to communicate their needs at different times and to cooperate for their mutual benefit.

Suzanne published her findings in *Nature* in 1997. The article was entitled "the Wood Wide Web". The article elicited a storm of reactions and criticisms both from scientists and the logging industry. But Suzanne knew her science was sound, and she was convinced that she was onto something important. Despite the controversy, she was offered a position in the forestry department at the University of British Columbia. There she was able to continue her research into the relationships between various species of trees and fungi.

One of her discoveries was that although the biggest trees nurture the other, smaller trees around them, they show a preference for their own

Right: Suzanne Simard 2018, sitting in a giant Mother Tree

offspring. Through the fungal network, they share their nutrients with other trees plugged into the network, but they somehow recognise their own seedlings and transfer more carbon to them. Trees actually mother their young! The social relationships in the forest are uncannily similar to those of human beings, in which we give the most care to our immediate families but also cultivate privileged relationships with members of the surrounding community.

Her work led her to consider indigenous cultures, which emphasise the bond between human beings and nature. Suzanne developed relationships with Coast Salish people, the aboriginal people living in the coastal region of British Columbia. She felt that her discoveries resonated with their views. The things she was discovering and trying to prove with the methods of western science were already known to indigenous people through centuries of observation and a respect for nature. Suzanne herself had intuited much of what she had discovered because of her own close connection to nature, but her ideas only gained credibility in western society through years of replicated, scientific studies.

The sense of this interconnectedness inspired her to expand her investigations beyond the web of tree species and fungi. She discovered that in the rings of Douglas fir, growing upland high above the rivers, there were traces of nitrogen derived from salmon. The salmon, a species sacred to the Coast Salish peoples, return from the ocean to journey upriver to where they were born in order to spawn. After they are born, the young salmon return to the ocean, while the adults remain in the river to die. **Bears carry salmon carcasses up into the forest to feast on them under the canopy of the Douglas fir. The remains decompose and seep into the fungal network of the mother trees and are redistributed through the forest community.**

Outside the scientific world, Suzanne's work gained more traction. She was invited to give TED talks; and published a book, *Finding the Mother Tree: Discovering the Wisdom of the Forest* in 2021. Her work partly inspired the "Tree of Souls" in the blockbuster film *Avatar,* directed by James Cameron.

She has even achieved more respect and recognition in the scientific community.

Suzanne and her research team undertook The Mother Tree Project, which aims to identify methods of harvesting and regenerating forests that would increase their resilience to climate change. The hope is that its findings will help to ensure the survival of our forests as well as establish a sustainable future for logging.

Twenty-three experimental sites, including Douglas fir forests, have been established in eight regions of British Columbia, each region lying in a different climate zone. The experimental sites measure roughly 20-25 hectares and are divided into 4–5-hectare plots. There are five experimental treatments, one for each plot. One is left untouched as a control, another is clearcut. The three remaining plots are subject to experimental conditions. In one, trees are cut down, leaving only the large mother trees, about 25 trees per hectare. In the other two plots, intact patches of trees and vegetation remain uncut around mother trees: 30% of the surface for one group and 60% for the other. A mix of tree seedlings is then replanted in the cut areas. The experiment is designed to continue for a hundred years, from which data will be assessed every few years. These experiments aim to lay the basis for a more resilient and sustainable forest management. Can we harvest wood from forests while maintaining them as functional ecosystems? Will the islands of intact ecosystems left between the logged areas help the new trees to grow and thrive?

As the knowledge that trees, like us, depend on their social networks becomes more widely known, perhaps we will stop clearcutting forests and instead harvest them in such a way that they will easily regenerate because of the underground ecology they require for their health. Suzanne's work will provide data to inform the forest industry in British Columbia and beyond. Her message has inspired others around the world.

Will we catch on in time to apply more sustainable practices to other forests around the world? We certainly can't wait for the final results in a hundred years before we take action to save the world's forests.

Wetland ecosystems

Wetlands are areas that are either covered or saturated with water, such as coastal ecosystems, floodplains, swamps, or lowland lakes and ponds. These areas, critical for replenishing the reserves of groundwater and aquifers, form when rainwater becomes trapped in a flat, low-lying area and gradually settles into the ground. The longer the water remains, the more water there is available to keep the landscape hydrated.

With the extremes of heat brought on by climate change, the best way to provide resilience to life on Earth is to have as much water as possible available for periods of hot, dry weather. Wetlands, lakes and ponds are nature's stock of water. However, water tables are being depleted through poor management. Natural ecosystems which help water to percolate and recharge the water table are being destroyed to make room for human development.

When it rains, some of the water runs off, taking organic material with it; some is absorbed by soil and plants; and some infiltrates the earth, trickling down until it reaches an impermeable layer. There it accumulates, building up pressure. It is this pressure that creates springs and keeps streams and rivers flowing in hot, dry periods. Roots of trees seek out these underground stores and draw them upward, providing water for other plants and microorganisms which dwell in the upper layer of the root zone. Evapotranspiration ensures cooling through clouds and rain. When the water table is too low, all of this stops. Streams and springs dry up, vegetation wilts and dies, the rains stop.

Wetlands can be restored if we manage the land around them in such a way to increase water infiltration. But we also need to create spaces in our managed landscapes which can retain and store water.

Marcus Dittrich: Building Water Retention Landscapes

Marcus Dittrich is a member of the Ecology Team at the Tamera Peace Research & Education Center in southern Portugal. Tamera is renowned for being a leading example of a water retention landscape. Marcus and the Ecology Team develop and manage the hydrology of the ecosystem on their 150-hectare land. The mission of Tamera is to research, develop and demonstrate lifestyles that allow people to live in harmony with each other and with Nature.

The community was established in Germany in the late 1970s by Dieter Duhm, Sabine Lichtenfels and Charly Rainer Ehrenpreis. They were trying to find solutions to the technological and environmental problems affecting the world at that time, by addressing core human values which had been eroded by competition, greed and violence. Today this ecovillage comprises about 170 full-time residents plus a flow of international visitors who participate in workshops and courses.

Marcus was born and grew up in East Germany, where he had the privilege of living close to a national park and wildlife reserve. He spent his childhood roaming the woods. Over time, he realised his life was privileged, that the distribution of wealth and power in the world is inequitable and that nature is exploited and degraded all over the globe. He decided not to go to university, which he felt perpetuated the forms of exploitation he had come to reject, and instead moved to Switzerland, where he studied spiritual and psychological healing, bodywork and martial arts. The ideas behind these practices inspired him to study ecology and permaculture. It was at this time that he met Sepp Holzer.

Sepp, who calls himself a "rebel farmer", taught permaculture, and developed methods for decentralised water retention, which he taught on his farm in Austria. Marcus was inspired by what he calls Sepp Holzer's "loud and unconventional voice". He, along with Sepp's other students, visited Sepp's farm and similar sites elsewhere in Europe. Sepp teaches his students to "read the book of nature". Rather than learn a method, they learned to observe and understand the basic principles of nature and create solutions adapted to specific environments. Marcus wanted to make his own contribution to improving the world. Sepp and others in his entourage told him about a project called Tamera.

Sepp became a mentor to the Ecology Team at Tamera in 2006, before Marcus arrived. At the time, the community was suffering from a water shortage. Sepp said the problem was not a lack of water; there is enough water if you know how to engage with it. Under his guidance, Tamera built its famous water retention pond in the centre of the community and put into place auxiliary methods of rainwater harvesting to maximise the infiltration capacity of the land they steward. Marcus joined Tamera in 2009. That year the water retention pond overflowed into the valley downstream, a cause for celebration and proof to the sceptics that retaining rainwater run-off could indeed shift the local hydrology.

Southern Portugal gets plenty of rain, nearly as much as Berlin, but it's very concentrated in the winter months. The rest of the year can be extremely hot and dry. Industrial methods of agriculture and degenerative forest management have caused a detrimental loss of topsoil, vegetation cover and biodiversity. With decreased infiltration capacity, the land in Portugal spiralled into a process of desertification. During the 1960s the country's dictatorship pressured farmers to abandon traditional methods of agriculture and turn to wheat production.

Traditional agriculture in southern Portugal had evolved to adapt to the climate and topography of the region, and was known as the "Montado agroforestry system". Large cork or holm oaks covered grasslands grazed by ruminant animals. There were also olive and other fruit trees, interspersed with some production of vegetables. The cash crops were cork, acorns, meat, olive oil and vegetables. The grass and trees protected the land from soil loss in the sloped hills and they helped to store water in the landscape. Annual production of grain, which depended on regular tilling and chemical inputs,

   Right: Marcus Dittrich working with children at Tamera

was completely unsuitable for the portuguese climate, especially in the hills, and it degraded the land, releasing its stored carbon and water.

Aside from the signature pond in the centre, the Tamera community has built smaller earthworks as well. There are swales, berms and ditches designed to capture water flowing down a slope. Mounds have been raised on the downhill side of the swales and stabilised by vegetation to soak up the water as it drains. There are check dams, made from natural materials like rocks, branches and vegetation, which stop the flow of water, forcing it to spread out across the landscape where it is slowly absorbed.

Over the years, the community has shifted its efforts from large earthworks, such as the central pond, to smaller projects all across the land. Planting the appropriate vegetation is a less obvious but important tactic in water retention. Thousands of trees have been planted on the slopes in an effort to recreate the original natural biome of the region. Context is important. Marcus points out that you need to understand where you are within the watershed. The appropriate actions will be different on the floodplain than in the headwaters, where, because of the higher elevation, the water runs

down faster. You also need to understand the local climatic variables. Tamera is at the headwaters of the larger basin of Rio Sado, where the upper slopes collect the rainwater. It naturally runs off the slopes, so the soil will be shallower and less fertile than soil in the floodplain. But if trees are planted on the slopes, their roots will stabilise the soil and absorb water. Anything that disturbs the soil will likely lead to erosion. The slopes have been planted with indigenous trees, restoring the landscape to its original condition. As a result, wildlife, including wild boar, has returned.

Other water harvesting methods include food production with living soil, practicing no-till gardening, and using mulches to build soil and retain water. The community produces only a portion of their food with their no-till gardens, fruit trees and olive grove; they do not live fully off the land. Dry, composting toilets minimise water use. The implementation of these methods has restored the surface hydrology. Shallow wells once dry are now full and serve their needs.

Yet the region remains deficient in groundwater recharge - a phenomenon seen in large areas of the Iberian Peninsula. The efforts of Tamera are insufficient to redress the hydrology of the whole region. Marcus knows an 80-year-old former shepherd who remembers that the creeks once flowed during the summer. Now they don't. Woody shrubs and trees are the only plants to remain green in the summer. The land is dry and brown until the rainy season returns. People dig bore holes 80 meters or more to draw water. Unfortunately, the neighbouring farms have not adopted the water harvesting practices of the Tamera community.

The latter is in a process of re-evaluation. Despite their success, the Tamera ecovillage has been unable to persuade the whole local community to adopt their practices. It is located in the municipality of Odemira, whose inhabitants come from families that have been here for generations, or from international citizens, as well as from the growing number of immigrant workers who flock to Portugal to enter Europe.

Right: The signature water retention ponds of the Tamera community have helped to restore the surface hydrology of the local region.

*On a sloped landscape, a good way to retain water is to plant perennials **along the natural contours of the land**. They soak up water and stabilise the soil, preventing erosion.*

The new generation at Tamera feel that not being rooted in the local community is a shortcoming and betrays the deeper values on which the village was founded. In order to rebuild the ecology at scale, humans need to come together. In a rural region that is not very populous, these disparate groups live without much integration. The next generation hope to become more integrated into the place where they live. Only when the tissue of the human cultures weave together, can the ecological values lived at Tamera seep into the practices of the community at large.

Valer Clark: Hydrating the land to bring back life

Valer Clark has worked for over 40 years to restore the desert lands of southern Arizona and northern Mexico. Valer says she was the most unlikely person to have taken on this work. She was born and raised in a wealthy family in New York City. As an artist and school teacher, she never imagined she would end up on a ranch in rural Arizona. But while she and her husband were visiting friends in the state, they decided on a whim to visit a ranch that was for sale in the Chiricahua Mountains. Her husband suggested that they place a low bid on the property just for fun. Valer agreed. To their astonishment, their offer was accepted and they became the owners of the El Coronado Ranch.

It is located at the intersection of several biomes, where the Chiricahua Desert meets the Sonoran Desert, and where the northern mountains meet the southern mountain range. The intersection of biomes is always where the greatest degree of biodiversity occurs, and it was certainly the case here. The couple's plan was to use the ranch as a holiday home, but as she spent more time there, Valer became fascinated with the landscape.

During the summer monsoons, she witnessed rivers of mud pouring off the mountains, carving out deep gullies and cracks and washing the soil away. This seemed like an aberration of nature, destroying itself in this way. How could it be normal for nature to create a landscape that worked like this? As Valer pondered this question, she concluded that this was not a natural phenomenon but the consequence of human intervention. The cause was Arizona's "three C's": copper, cattle and cotton.

When settlers arrived in the 1800's, they found tall grasses and quickly brought in cattle and sheep to graze it. Following the discovery of copper in southeast Arizona, they clearcut trees in the Chiricahua Mountains in order to have fuel to process the copper. As the forests were turned into grasslands, ranchers brought more grazing animals up into the mountains. They cleared land in the valleys to grow cotton. By the time Valer and her husband arrived, local residents assumed that dry conditions and August monsoons were normal; nobody remembered it being otherwise.

One day, Valer and her husband noticed a line of stones on the ground, which created an obstruction in the pathway where the water would run off. This had caused some silt to collect in that spot, out of which grass was growing. It became clear to Valer that the unnatural situation was caused because vegetation had been removed from the landscape.

By chance, a group of Mexican men came by looking for work. They said that, where they came from, they built trincheras — rock structures on the hillsides that stopped earth and water from running down. They planted maize on these structures, taking advantage of the water and organic matter that had collected in them. Furthermore, having several of these trincheras together stabilises a slope. Valer hired the Mexican crew, which set to work building thousands of trincheras over the mountain side. The crew returned each year to add more trincheras, covering all 1800 acres of the El Coronado ranch, plus surrounding land which was under lease from the US Forest Service. Landowners in Arizona pay a tax for federal land reserves, and ranchers are entitled to graze their cattle on them. Over the next 20 years, the team of workers took the liberty of constructing these trincheras on 16,000 acres of Forest Service land.

Trincheras were constructed along the gullies, creeks and fissures caused by the fast-flowing water. The trincheras are not meant to stop the flow, just slow it down so that silt and debris can accumulate and create a habitat for plant life. After 40 years of patient labor over this vast area and despite extreme flooding that has destroyed some of the trincheras, ground reserves of water have increased, vegetation is growing, and creeks and rivers are flowing year-round for the most part. Laura Norman, a scientist from the University of Arizona, who evaluated the impact of the trincheras, calculated that areas where a trinchera had been built contained 28% more moisture than nearby areas without one. Even without scientific confirmation, the difference is apparent to the eye: these areas are green and full of life.

Once the streams were running again, Valer met with Dr Wendall Minckley, a leading expert on fish, who taught at Arizona State University. Dr Minckley suggested that she stock the streams with a species of small native fish capable of surviving dry periods. The species was nearly extinct because of the loss of its natural habitat. Together, Valer and Dr Minckley stocked the streams of the El Coronado Ranch. The fish survived and bred. Dr Minckley had been working with government and academic agencies and was dissatisfied with the progress of his work. He offered to train Valer to restock native fish. Valer joked that it was probably because he thought that she, not being a scientist, would follow his advice without question, rather than form her own views. In any case, she was happy to have been responsible for the reintroduction of a native species to its original habitat.

Encouraged by their success, Dr Minckley suggested that Valer and her husband expand their stewardship into North Mexico to include a particular stream that was a critical habitat for several species of endangered fish. He said it was probably the most important stream in the northern hemisphere. They bought a large property, known as the San Bernadino, situated along the border. The stream ran through this property, which at one time had been a huge wetland. The Mexicans called it a ciénaga or "hundred waters". The area had suffered from the same destructive practices as those that Valer had encountered in Arizona, and the consequences were the same. With the soil unable to retain rainfall, it ran off the land in torrents, cutting deep gullies into the land. They were 20 to 30 feet below the surface, draining the wetland. Trincheras would be ineffective here but something had to be done as the marshland no longer retained water throughout the year. Migrating birds and animals no longer stopped here. An area of rich biodiversity was slowly dying.

Valer's crew turned to a different technique. They made wire baskets, filled them with rocks and fixed them to the banks. The structures were called gabions, or "baskets of rocks", designed to hold the rocks in place. They were told that it couldn't be done, and indeed, many of the gabions broke

against the force of the water. But again, according to Valer, even if they broke, they had a beneficial effect. The crumbled gabions retained enough organic matter that soil accumulated. When the streams overflow, because of the soil, they hydrate the floodplain instead of cutting channels in the dry land and draining away. Vegetation is growing back. Trees were planted along the banks, using a method called pole planting. Cuttings of willow and cottonwood are placed in the soil on the banks. They take root, soaking up the water and stabilising the banks. Once they are well established the trees propagate naturally.

After two decades of efforts by Valer, her husband and their team, the land is becoming green again. Water is being retained by the landscape, even in areas lying several kilometres upstream and downstream from their properties. Twenty years ago, there was scarcely a living creature in the area. Today there are lizards and snakes, birds, fish, racoons, bobcats, coatimundis, bears, even coyotes, and recently jaguars and Mexican wolves were sighted. The abundant biodiversity of this ecosystem is flourishing again. I asked Valer if they actually do ranching because she hadn't once mentioned cattle, although their properties were called ranches. Yes, she told me, there are cattle, and they are one of the tools of the overall restoration. They are excluded from areas where tender vegetation has begun to grow, and they are managed in tight herds, which are moved every four hours and confined with moveable electric fencing. Their manure and urine are valuable in re-greening the land. She expects that some cows may be lost to the returning predators, but that tightly packed herds protect themselves, just as they had in the wild. Some losses, however, are inevitable if nature is to maintain a balance.

Valer is now in her eighties, and she wants to be sure that the 140,000 acres of recovering habitat she stewards in Mexico will be protected into perpetuity. Since 2005, she has adopted various legal measures to ensure her legacy, including the creation of a family foundation. Valer's daughter, Valerie, is on the board of directors. Valer also set up the non-profit Cuenca

Left: Valer Clark on the El Coronado Ranch, in the Chiricahua Mountains.

los Ojos, ("Watershed of the Springs") with a branch in Mexico and another in the US. Its mission is to protect, restore and rewild the biodiversity of the US-Mexico Borderlands. The words have a double meaning: cuenca ("eye socket") and ojos ("eyes"), which suggests not only the physical aspect of restoration but the understanding and vision needed to act.

The family hopes to secure additional funding to expand the project, permit individuals to make charitable donations of land to the trust, and offer programmes to train others in the effective management of their land to restore the watersheds. One issue that the organisation faces is that their territory lies along the US-Mexican border where the US government has constructed an enormous wall to control illegal immigration. The wall with its glaring lights has disrupted the movement of wildlife. Migrating birds returning from South America to North America are blinded by the light and get lost, while mammals such as bears and jaguars have lost much of their territories. Valer has seen animals struggling to get through the wall.

It is painful to see the harm caused by decision makers who are unaware of the travel corridors and territories that wildlife needs to sustain their populations and to migrate. Ecosystems do not respect arbitrary international boundary lines. Not only animals, but rivers and streams need to cross these lines to respect the integrity of their ecosystem functions.

There is much work to be done still before we learn to recognize that our wellbeing is embedded in natural ecosystems and we must learn to understand, respect and live within them.

Left: This trinchera is filling up with earth and grasses.

In the wetlands of northern Mexico, the crew built "gabions", wire structures that hold the rocks in place to slow the flow of water and collect sediment. Soil builds up in these places and trees stabilise the banks, so that the water hydrates the land and keeps the wetland habitat alive year-round. Now the water in this wash (a dry riverbed or seasonally active creek) is permanent. But it took many years to accomplish.

Chapter Three
The World Is Round: Big Patterns of Weather

So far, we have focused on the green water cycle, the movement of water in various regions and its interactions with the earth and the atmosphere. However, there are larger movements of the water cycle which have an impact on weather and climate. How we manage the land in an area has an impact on how resilient that area will be in extreme conditions. But other weather patterns, such as the large movements of atmospheric moisture, governed by the winds, determine how much heat and moisture move between the equator and the poles, and where.

Ultimately the Earth is one big ecosystem, which some call "Gaia". Through her movements, every corner of the planet is connected. The larger weather patterns affect us all. As the polar ice caps melt and the tropical forests are lost, this has impacts on the entire global weather system.

Many weather patterns are regulated by the Earth's rotation and its orbit around the Sun. Although all life depends on the Sun's energy, the same amount is not delivered to every place as our planet is tilted on its axis. 2.5 times more solar energy reaches the equator than the poles, creating a denser atmosphere. The Earth's rotation as it orbits the Sun causes the patterns of wind and ocean tides which distribute the Sun's heat. If Earth were stationary, the equator would be much hotter and the poles much colder.

Water evaporates when heated, converting solar energy to latent heat. The water becomes vapour, rises until it reaches a cooler layer of the atmosphere, at which point it condenses, releasing the heat at a higher latitude. This is how most heat is distributed. Water moderates temperatures, and winds move the streams of air and water vapour from the equator to the poles in a complex pattern of movement.

The roundness of the Earth means that the planet spins faster at its equator than at its poles. If you were standing at one of the poles, you would turn very slowly, but if you were standing on the equator, you would travel in a much larger circle. Because the Earth rotates, moving objects in the Northern Hemisphere deviate clockwise (toward the North Pole) and in the Southern Hemisphere counterclockwise (towards the South Pole). This is called the Coriolis effect.

Each hemisphere has three climate zones, governed by the movement of air currents which transfer heat around the globe. Warm air rises and cool air sinks, which in combination with the Coriolis effect, determines the movement of the air. At the equator, which receives the most heat, a pattern of circulation called the Hadley cell drives warm air upwards and away from the equator. When this air, now much cooler, finally sinks again it flows back down towards the equator. In the polar regions air flows down along the surface of the Earth until it reaches warmer air and then it rises as it warms and flows back to the poles. Between the Hadley cell at the equator and the Polar cells at the poles, lies a temperate zone affected by the movement of these two cells. It is called the Ferrel cell. It moves in the opposite direction of the Hadley and Polar cells. Where their air flows meet, the clash of colder and warmer air causes turbulence. You'll notice that whenever the weather changes abruptly from warm to cold, there is a lot of wind and sometimes a storm.

In each hemisphere there are two jet streams, one at 30° north and 30° south, where the Hadley cell meets the Ferrel cell, and a second, lying 60° north and 60° south, where the Polar cell meets the Ferrel cell. The polar jet stream runs

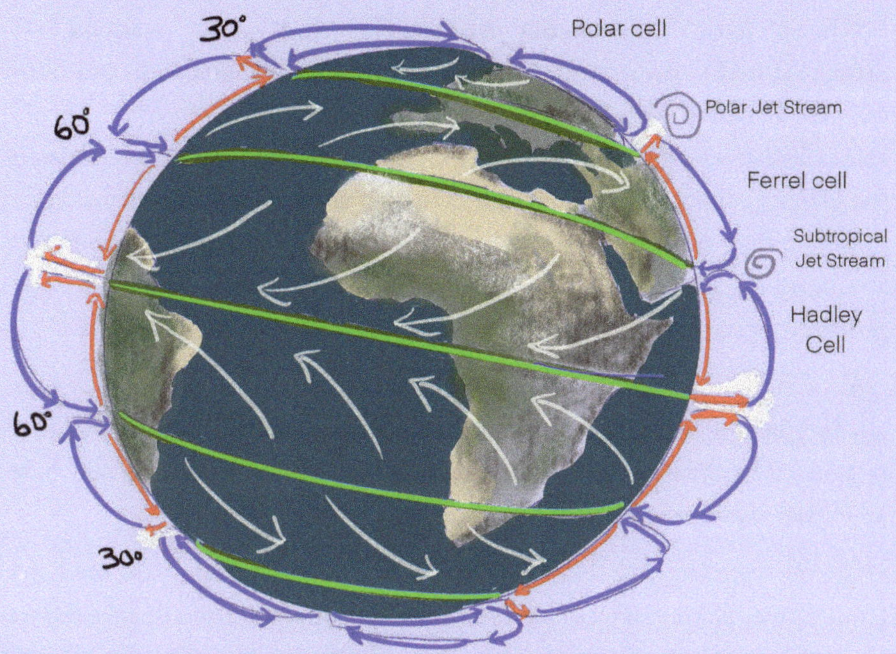

The rotation of the Earth circulates air. The equatorial zone receives the most solar radiation. Warm air rises and as it cools in the subtropical zones, it sinks, creating the Hadley cell and the subtropical jet stream. From the poles cold air flows over the surface and rises when it meets warmer air, creating the Polar cell and the polar jet stream. The Ferrel cell in the temperate zones is driven by the movements of the Polar and Hadley cells. This air circulation determines the typical weather patterns in each zone.

in the higher atmosphere, and the meeting of very cold and warm air creates strong winds, especially in winter when the contrast of air mass temperatures is greatest. The subtropical jet stream, arising from the meeting of the Ferrel and Hadley cells, creates lighter winds at a higher altitude. Jet streams always move from west to east in a wave-like pattern, driven by the friction between cold and warm masses of air. The movement of jet streams causes changes in the weather.

When air from the Ferrel cell meets the Polar cell, both rise, creating a low-pressure system. Low pressure systems are unstable, bringing cold and rainy weather, which is why countries at these latitudes tend to have wetter and more variable weather. On the other hand, when air from the Ferrel cell meets the Hadley cell, both air masses fall, creating a high-pressure system. High pressure brings dry and clear weather as a rule and this is the reason countries at around the 30° latitude have drier weather. These areas are warmer and generally receive less rainfall. This is where most deserts are located.

All of these air movements are caused by the shape and rotation of the Earth, not by global warming. This explains why the water cycle in one region may be affected by the weather in another, and why land use in one area may have an impact elsewhere.

Global warming affects the water cycle, and the water cycle mediates temperature. Warmer temperatures cause more evaporation and therefore more water vapour enters the atmosphere. As temperatures continue to rise, the ice at the poles melts, reducing the area that reflects heat away from the Earth. The drier the land becomes, the less hospitable it is to plant life. As a result, there is less evapotranspiration, more sensible heat remains trapped in the Earth's surface rather than rising up in the atmosphere, and air circulation is disrupted.

Oceans

The Earth is a unique planet because of the presence of liquid water. Oceans cover 71% of its surface. They absorb most of the carbon from the carbon cycle and most of the heat from the Sun. Unlike air or earth, which release heat quickly, water cools slowly, buffering the Earth from sudden shifts of temperature. This is why coasts are cooler in summer and milder in winter.

Oceans act as giant conveyor belts, moving currents of cold or warm water between the equator and the poles. Along the equator the oceans absorb massive amounts of solar radiation and also lose an enormous amount of water through evaporation. In the Southern Hemisphere, ocean currents are not obstructed by land masses, so they move smoothly around the Antarctica continent. In the Northern Hemisphere, however, the abundant land masses cause ocean currents to deviate from their course. Water warmed on the surface of the Atlantic Ocean flows upwards to Greenland and Northern Europe. Once it cools in the polar region, it sinks back down and flows toward the South Pole as a deep current along the ocean floor. Europe would not benefit from a temperate climate if the ocean did not transport this heat northwards.

This circulation in the Atlantic Ocean is called the Atlantic Meridional Overturning Circulation, or AMOC, and according to climate scientists, climate change is causing it to slow down. Like the movement of air currents, ocean currents are generated when warm water cools as it nears the poles. Since colder water is denser than warm, it sinks. As Earth gets warmer at the poles, it slows this turnover. Another factor leading to this sluggish turnover is that salt water is denser than fresh, and melted ice flows dilute the ocean's salinity when they flow into the sea.

The deep water current rises up again in the Pacific Ocean, carrying the equatorial heat from the coast of South America towards the Australian continent and then back around the African continent to the Atlantic. Changes in the ocean's surface temperatures can have huge impacts around the world. For example, the El Niño Southern Oscillation (ENSO) is a

Left: This thermal image shows that when the sunlight falls on bare compacted ground, temperatures may be as high as 47°C, whereas in vegetated and shaded areas it ranges from 25-35°C.

The ocean currents transport heat around the globe.

The red arrows show warm surface currents, and the blue arrows deep cold currents, which run close to the ocean floor.

phenomenon which occurs every few years depending on a variation of temperature in the equatorial Pacific Ocean off the coast of South America. If surface temperatures in this region remain higher than normal for an extended period during the autumn, the warm water creates a low-pressure area which blocks the normal flow of the westerly trade winds. Instead of warm moist air being carried towards the western Pacific by the trade winds as it normally would, it backs up, causing stormy weather and rains along the coast of South America and up into North America. Without the usual moisture being transported westwards, Australia, Indonesia and Asia along the western Pacific experience exceptional heat and droughts. The warm ocean surface temperature in the eastern Pacific off the South America coast suppresses the upwelling of the current of colder, deep waters. Without this upwelling, nutrients necessary to nourish the marine ecosystem along these coasts never reach them, and they cannot support the usual abundance of fish and marine life, causing hardship to the people of Ecuador and Peru, who depend on fishing for their livelihood. El Niño can even have an impact on countries on the other side of the Atlantic side, reducing the severity of winter temperatures and storms that form in the Atlantic come autumn.

El Niño's counterpart, La Niña, occurs when the equatorial surface of the Pacific remains cooler than normal for an extended period. The moisture-laden westerly winds blow more strongly towards Australia, Indonesia and Asia, bringing heavy rains and flooding to those countries. La Niña can also exacerbate tropical storms in the Atlantic. El Niño and La Niña are not caused by climate change, but like other normal climate phenomena they have become more extreme as the planet warms. These extremes of weather, the general rise in sea temperatures and slower circulation of ocean currents are a direct result of climate change.

The ocean occupies a huge proportion of the planet but scientists are only beginning to understand the function and impact of marine ecosystems on the climate. Just as the green water cycle is supported by terrestrial plants, tiny photosynthetic marine plants, called phytoplankton, have an enormous impact on the ocean's carbon and water cycles. Phytoplankton live in the warm top layer of the ocean, where they have access to light. Because they are photosynthetic, they, like other plants, absorb carbon from the atmosphere and convert it into biomass. Phytoplankton are the basis of the aquatic food web. Marine herbivores consume it and larger marine creatures consume those. The movements of these marine creatures distribute the nutrients and biomass throughout the deeper layers of ocean where light does not penetrate. They produce waste, they die, and eventually sink to the ocean floor, adding to its biomass. The ocean is the largest carbon sink on the planet, containing fifty times more carbon than the atmosphere.

The circulation of warm surface waters, where phytoplankton live, with the ocean floor where nutrients have been deposited as sediment, is crucial to both climate regulation and the cycle of life. Life requires both the photosynthetic activity of phytoplankton and the circulation of nutrients from the depths. If the cycle is broken and nutrients fail to rise, marine life is deprived of food. The cycle is driven both by ocean currents resulting from the rotation of the Earth and by the movement of marine life between the ocean's layers. Whales play a particularly important role, as they travel between the surface and the depths, as well as between equatorial waters and the poles. In addition, a recent study has shown that their surface excretions favor the production of plankton, so they play a role in the oceans which is similar to that of ruminants in grasslands.

Phytoplankton also have an effect on climate through the creation of clouds, like their larger terrestrial relatives, green plants. This idea was first put forward by British chemist James Lovelock, the creator of the Gaia theory. It posits Earth as a living organism made up of complex, interconnected ecosystems which are self-regulating. Life creates the conditions to keep itself in balance. Too much solar radiation disturbs phytoplankton. Lovelock was the first to notice that phytoplankton react to UV light by producing a substance called dimethylsulfoniopropionate (DMSP), which thickens the walls of their cells as a protection. When DMSP is dissolved in seawater it breaks down to dimethylsulfide (DMS). This substance quickly breaks down into sulphur compounds, which volatilise to become aerosols. These tiny aerosol particles are cloud condensation nuclei, and they cause water vapour to condense into clouds. Clouds block the Sun's rays and thereby protect the phytoplankton. They also stop heat from being absorbed by the dark ocean surface. Since Lovelock proposed these ideas, they have been validated in numerous studies, which confirm that phytoplankton produce DMS, thus helping to control sea surface temperatures.

Phytoplankton need both sunlight and nutrients from the ocean floor to carry out this task, and they are a keystone in its complex self-regulating ecosystem.

The ocean absorbs more heat and stores more carbon than any other ecosystem on the planet, buffering the effects of climate change. But we can't rely on it to save us indefinitely. At the interface between the seawater and the air, carbon dioxide enters the salty ocean waters. The salt dissolves the bonds between carbon and oxygen, creating carbonic acid, used by shellfish and corals to build their structure. But as the levels of CO_2 increase, carbonic acid breaks apart, releasing bicarbonate and hydrogen ions. This causes ocean acidification, which is a lowering of the pH, the measure of hydrogen ions in a solution - in this case seawater. The more CO_2 is absorbed by the ocean, the more it becomes acidic. As with all ecosystems, marine life has evolved to function at certain temperatures and pH levels. As the surface of the water warms and becomes more acidic, phytoplankton cannot reproduce, shellfish and coral cannot make their shells. These rank at the bottom of the marine food chain, and their loss threatens the larger marine life that depend on them. **Without these marine ecosystems, the ocean will stop absorbing our excess CO_2, and the ocean conveyor belt will slow down.**

Extreme Weather

Millan Millan was a Spanish scientist who studied atmospheric physics and meteorology at the University of Toronto. He worked for many years in Canada before he was invited to head a study for the European Commission to investigate the Mediterranean Basin, specifically to understand the reduction in summer precipitation. Millan was active in the early days of the climate change conversation. He was at MIT (Massachusetts Institute of Technology) when world leaders first met to discuss the possibility of human induced climate change. He remembers that, at the time, the conversations focused on two drivers of climate change. One was greenhouse gasses, caused by burning fossil fuels, the second was land use change. Humans had transformed the landscape by clearing away natural ecosystems and replacing them with farms, mines, roads and buildings. Millan's study of weather changes in the Mediterranean Basin convinced him that land use change is the more significant cause. After he completed his study for the European Commission and the results were being shared around the world, he began to recognise similar patterns elsewhere.

Until a few centuries ago, there were forests in the mountains of France and Italy and marshes in the coastal areas of the Mediterranean. In summer, winds blew in each morning from the sea and as they rose, they warmed and moved up the mountains. At an elevation of about 700 metres, cooler temperatures caused moisture carried by the breezes to condensate, bringing rainfall in the afternoon. This was a regular feature of the Mediterranean climate. But over time the summer rainfall decreased. Forests had been cleared, and marshes filled. In the 1970s development accelerated: more buildings, oil refineries, roads, and farms. Ozone began to accumulate in the atmosphere of the region. That was when the European Commission contacted Millan Millan.

He concluded that the lack of summer rains had caused the accumulation of ozone, not being washed from the atmosphere. For it to rain, enough moisture must be in the air when it cools, causing the water to condensate.

The sea breezes were bringing in about 14 grams of water per kilogram of air, but in order to compensate for the heating that occurs as air flows inland, it would need to be carrying 21 grams/kilo to create precipitation as it rose to a higher and cooler altitude.

Sea breezes normally pick up extra water from evapotranspiration as they travel over marshlands and vegetation. Historically in summer in the Mediterranean Basin, sufficiently moist air would cause it to rain almost every day in the afternoon, hydrating the landscape. The same water would be recycled constantly, because the daily rains supplied enough water to the soil and vegetation to increase the moisture content of the air blowing inland the following day. The vegetation and the hydrated soil contributed to the regular rainfall and perpetuated a virtuous cycle of comfortable temperatures and clean air.

But today over a drier landscape, the sea breeze carries only 14 grams/kilo of moisture, rather than 21 grams/kilo. This is not enough to condensate as rain, so it accumulates in the atmosphere. In autumn and winter, when the air over the land cools faster than the air over the sea, the moisture-laden air which has accumulated over the sea is pulled back inland. This often causes torrential rains and flooding in parts of Europe. Which areas of Europe will be affected depend on various factors, particularly wind direction. When too much rain falls in area with poor, dry soil and insufficient vegetation, it doesn't soak in to feed the groundwater, instead it causes soil erosion and flooding.

Millan says there are series of tipping points which transform this virtuous cycle to a vicious cycle. The first tipping point is the loss of soil moisture from the removal of vegetation. This leads to a loss of the evapotranspiration necessary to trigger rainfall. The ground is hotter and dryer so the rains no longer fall in summer. Dry land means a greater likelihood of wildfires, which desiccate the land further. The build-up of moisture that normally falls as rain in the summer eventually returns when temperatures drop in autumn, bringing heavy rain and storms.

The lack of water absorption caused by reduced vegetation as well as a lack of soil porosity from drying and compaction leads to the second tipping point, loss of soil. Heavy rains wash it away. There are mudslides. Fertility is lost, the landscape becomes bare sand and rock. It accumulates even more sensible heat when it no longer holds enough water or vegetation for evapotranspiration.

Millan says that "Water begets water, soil is the womb, and vegetation is the midwife". He sees the same pattern emerging in many other parts of the world. The southwest United States and Australia are experiencing a similar cycle of droughts and wildfires followed by intense rains and mudslides. Longer and more severe droughts are not the opposite phenomenon of intense floods; they are the other side of the coin.

Sadly, Professor Millan passed away in January of 2024. He was disappointed that the second driver of climate change, the loss of natural infrastructure, was absent in the climate conversation. He believed, along with many other scientists, that the cycle of drought, fire and floods brought on by climate change can be more quickly remediated by restoring the soil, vegetation and water cycles in arid areas than by waiting for governments to reduce global emissions.

Pakistan

In the summer of 2022, Pakistan suffered the worst flood in its history. In the preceding April and May, temperatures exceeded 40° C, making it the hottest place on Earth. Soaring temperatures caused an abnormal melting of the country's northern glaciers, with the result that the rivers and drainage ditches were already full when the monsoon season arrived. The United Nations Secretary-General Antonio Guterres called it "a monsoon on steroids". Infrastructure was washed away; crops and livestock were destroyed. Over two thousand people were swept to their deaths and

33 million others were left homeless. A third of the country lay underwater. The government estimated damages at 30 billion dollars.

The prevailing story is that the disaster was caused by carbon-induced warming, and that although Pakistan generates only 1% of global GHG emissions, it suffers the worst consequences of climate change. The disaster helped to raise awareness of the unequal impact of climate change on different countries. Developed nations have been called upon to assist poorer nations financially to deal with these impacts. This is a positive step, but if we don't understand the full genesis of these disasters, funding may be spent ineffectively. Fossil fuel emissions alone did not cause the disaster in Pakistan.

Pakistan does receive an abnormal increase in rainfall during the monsoon season, partly due to global warming, but the country is also the victim of corrupt and unchecked exploitation of natural resources. In 2011, the UN listed Pakistan as the most deforested country in Asia. Every year 27,000 hectares of natural forests are felled for timber, or to make land available for farming and urbanisation. The unprotected soils have lost much of their capacity to absorb water. Forest clearing around the Indus River has disrupted the riparian zone and altered the hydrological system. The infrastructure of canals, dams and reservoirs has been neglected, left to accumulate silt and clog up, and illegal construction occurs on floodplains. Like many poor countries, Pakistan has succumbed to the pressures of globalisation in the exploitation of its natural resources.

Unless the fundamental problem of pillaging ecosystems which help the land to deal with excessive stress is addressed, Pakistan and other countries like it will fare no better in the future.

Vijay Kumar: systemic transformation in India

Neighbouring India suffers similar problems, such as crippling heatwaves and disrupted monsoon rains. The amount of groundwater has declined dangerously in many parts of the country. India also suffers from soil degradation, largely due to overuse of agricultural chemicals. The high cost of agricultural inputs has forced many farmers into a downward spiral of debt, and an increase in suicides. The state of Andhra Pradesh has embarked on a systemic change in policy, which could signal a major shift both for farmers and the local ecology. 70% of India's population are smallholder farmers who rely on agriculture for their livelihood.

T. Vijay Kumar began as a career civil servant in the Indian government, and now works with the state government of Andhra Pradesh. He leads an ambitious project called Natural Farming Food System for People and the Planet, which aims to convert farmers to adopt an agroecological approach by 2027. Its genesis goes back to 1993 when the United Nations Development Programme (UNDP) published a report concerning the causes of endemic poverty in six South Asian countries. It concluded that social organisation was a key factor. Women in particular suffer from extreme poverty, with few resources to help better their situation.

As a result, the Indian government launched a five-year pilot programme led by Vijay, which addressed the needs of 200,000 rural women. Part of it involved organising support groups. Poverty is a multidimensional issue. Poor women particularly, are disempowered through class and gender discrimination. When the programme ended, Vijay continued this work, this time under the aegis of the state government of Andhra Pradesh, heading an initiative called Society for the Elimination of Rural Poverty. He led this programme for ten years, helping some 11.5 million women through local self-help networks. Each village hosted a group of 10-15 women, which were connected to similar groups in other villages, then participated in larger regional meetings, and in turn at the district level. Each woman gained the support and solidarity of hundreds of others like her, struggling with poverty, domestic violence, and healthcare.

Besides the groundswell of support and solidarity, the programme also provided legal aid, women's shelters for victims of domestic violence, microfinance loans to help them start their own small businesses, education and technical support to learn skills that would allow them to earn a livelihood. All issues were addressed collectively. Women were empowered to create better lives for themselves and their communities. Their strengths, interests and talents were cultivated, allowing them to become leaders in their communities.

Two women in the programme, concerned by the health impact of pesticides, proposed in 2004 a project to help smallholder farmers reduce their dependence on chemicals in pest management. In the early 2000s the land in the state of Andhra Pradesh was fairly barren. Heatwaves were more frequent, monsoon seasons shorter and more unpredictable, tree cover was disappearing along with biodiversity, and water was growing scarce. The average family, which farmed about a hectare of land, could barely afford the inputs to eke out a living. The women wanted to change the situation to assure food security in the region.

The success of this model in Andhra Pradesh inspired India's central government to try to replicate it in other states. In 2011, Vijay laid out his vision of scaling up the model to reach 100 million women throughout India. As of 2022, 80 million women in 39 states were part of this national programme, which receives funding from both the state and the central government.

In 2015, Vijay became the head of the Department of Agriculture in Andhra Pradesh. Agricultural policies in all states support the use of chemicals, but Vijay was determined to launch a pilot project in agroecology which would address both poverty and climate change. He was confident that the methods he had developed with the women's networks would instigate a dramatic change in agriculture practice and provide food security and

Right: Mr. Vijay Kumar with farmers in Anantapur district, explaining the importance of 365 days of green cover to a delegation from Africa.

climate resilience across Andhra Pradesh. Vijay's colleagues were skeptical, but when Vijay challenged them to come up with a better idea, and no one could, his plan was adopted.

The methods chosen were based on the teachings of a man named Subhash Palekar, called Zero Budget Natural Farming. It was centred on rebuilding the soil microbiome with locally available ingredients, principally cow urine and dung. The excrement from one cow could supply enough inoculant to treat 20 acres of land. To start the process, seeds were coated with a mixture of urine, dung and lime. The whole community, through the efforts of the women's networks, would get together to treat the seeds before the planting season. Next, inoculants would be prepared, using urine and dung, with the addition of good soil, lentil flour and sugar, which would be spread over the field before planting. After that, the ground would be covered with crop residue to protect the microbial life from the Sun. With these methods, the soil structure would build up rapidly, permitting it to contain more air and absorb water vapour.

The farmers would plant annual field crops between rows of trees and bushes, depending on their height. With this method, called multi-storey planting, trees provide shade, help regulate temperature and reduce the loss of moisture. Species grown in the same area complement one another, filling different niches: some plants grow tall, others spread out along the ground, another thrives in shade. The main crops included millet, sorghum, rice, groundnuts, sesame, turmeric and ginger. Agroforestry species included cashew, mango, guava, tamarind and pomegranate. Once the seeds sprout, biological extracts from plants are used for pest management. None of the inputs come from outside the village, so there is zero cost for them once the system is operating. Saving the seeds for the next season is an integral part of this closed loop system.

The project began with a 10-day training programme led by Subhash Palekar. Some 8,000 farmers took part. Its ultimate goal is for farmers and the community to teach others. This is called the "saturation approach".

In the first year, the women are to recruit 15% of village farmers, many of whom are at first skeptical about abandoning chemical inputs. After the first year, they assess the results. In the second year the aim is to recruit 50% of farmers and 80% by the third year. An average village is home to 400 farming families. It will take about 3-6 years for them to adopt all the practices and apply them to the entire farm. The aim is that, after six years, the village will no longer require assistance and will be declared a bio-village.

The project uses the same model as the women's networks, choosing farmers who are the "community diamonds", having the talent, spunk and charisma to become leaders and teachers. In this grassroots manner, habits and behaviours can be changed through local influencers. A growing number of these champion farmers are being showcased as examples as the programme is introduced in more villages. The work is managed and monitored by the women's networks.

As of 2019, 28.000 farmers had been trained and 5,900 champion farmers were disseminating the knowledge to more villages. Farmers reported yield increases of 14-30% and net profits increase by 46-200%, thanks to the combination of better yields and lower input costs. Research and innovation continue in the project; Vijay believes that the methods need to be continuously adapted, including collaboration with researchers from universities and agroforestry experts. In 2018, Vijay came across a video by Walter Jehne, a well-known microbiologist and climate researcher, who pointed out that desert plants draw water directly from the air. He also claims that 50% of the water that dry land plants use comes not from rainfall but from the air. Vijay said it blew his mind when he heard this. It gave him the idea to try dry seeding. Farmers had been telling him that with the soil improvement, their crops fared better in droughts and required only half the irrigation as before. As Vijay says, after hearing Walter, he put 2+2 together and got 10.

In India and in many countries which depend on monsoon rains, the growing season begins when the rains come. Vijay and his research team

Villagers use readily available local materials to prepare natural soil amendments and seed coatings.

decided to try planting in the dry season, on a plot where the soil biology had been improved. To everyone's surprise the seeds germinated, and the plants grew. The plot was an oasis of greenery in a dry, bare field. Walter found that it takes a small amount of water for germination and that biologically enhanced soil contains water. Once a plant puts forth leaves, they draw moisture from the air.

This discovery was a game changer. Whereas traditionally farmers could only produce one crop per year, depending on the monsoons, they could now grow crops year-round. By using dung-based seed pelleting coating and multi-storey planting, they could store enough water in the ecosystem to grow five times the number of crops. In 2022, 600,000 farmers used pre-monsoon dry seeding. Farmers can now harvest 10-15 different crops from one plot over the course of a year.

The ground is never tilled and kept covered with vegetation. Some farmers use 30 varieties of plants as cover crop following the harvest. Cows grazing on them improve the quality of their milk. Every few weeks, year-round, there is something to harvest. The research and experimentation continue. Going forward the project will focus more on trees to plant between the fields to improve the microbial biodiversity and add even more harvestable crops.

As of 2024, nearly 10% of all the farmers in Andhra Pradesh took part in the programme. Vijay feels this is the tipping point. By increasing the number of champion farmers each year and showcasing their remarkable successes, he is confident of achieving his goal of converting the whole state of Andhra Pradesh into natural, climate resilient farming by 2027. After that, perhaps, all of India will follow suit.

The Amazon

The Amazon River basin is one of the most important ecologies on the planet. It has been referred to as the "lungs of the Earth". Antonio Nobre, a climate scientist who has been studying the Amazon for over 20 years, says that it should be thought of more accurately as the heart of Earth and also the liver because it circulates water, like blood, and purifies the air. The Amazon ecosystem is so important not only for South America but for the weather circulation patterns of the entire planet.

We saw that the equatorial region receives more solar energy than anywhere else. It is home to many tropical forests, located in Africa, Malaysia, Indonesia, Papua New Guinea and on islands in the Indian Ocean, but the Amazon Rainforest is the largest. Rainforest is the natural habitat for these regions, yet it is quickly being destroyed. If you look at a satellite map of moisture patterns over the Earth, you will see that the bulk of water vapour emanates from equatorial rainforests. They generate most of the world's continental rainfall and moderate temperatures in the warmest regions.

Most of the planet's heat comes from the equator and is redistributed towards the poles, and the same is true of water vapour. It stands to reason that if it is hotter and drier at the equator, it will be hotter and drier everywhere.

Antonio Nobre calls the Amazon "the splendid cradle of life". If we think of water as the blood of Earth, then rivers would be the veins. They drain water from the landscape and carry it to the oceans. What brings life-giving blood to the landscape? Rivers of evaporated moisture generated by the blue sea, says Nobre, and they are pumped over the green ocean - forests. From satellite images we see a pattern of moisture pulsing like a heartbeat.

In the Amazon there are 600 billion trees. Each tree can pump up to a thousand litres of moisture per day, five times more than an adult tree of the more temperate regions. That makes 20 billion metric tonnes of moisture released every day into the atmosphere, more than the Amazon River, the largest river on Earth, pouring 17 billion metric tonnes of water per day into the ocean. In total it is 37 billion metric tonnes of latent heat which is not heating the Earth's surface, and it is equivalent to 50,000 times more power than is generated by the largest hydroelectric dam. This energy re-circulates primarily over the Amazon, not over ocean, because the volatile aerosols emitted by the trees cause it to rain, purifying the air of pollutants. The air over the rainforest is the purest on the planet.

Antonio Nobre describes a meeting he had many years ago with Davi Kopenawa, a shaman and spokesperson for the Yanomami tribe. They were at a climate conference. Davi Kopenawa said, **"Doesn't the white man know that if he destroys the forest there will be no more rain? And if there is no rain, there will be nothing to eat or drink?"** He stated what Nobre and thousands of scientists had concluded after twenty years of study. Afterwards, Nobre puzzled over how this indigenous Yanomami elder had discovered this fact. When the two crossed paths a few years later, Nobre asked Davi Kopenawa how he came to this conclusion. He said simply, "The spirit of the forest told us."

Antonio Nobre is a proponent of the biotic pump theory proposed by Victor Gorshkov and Anastassia Makarieva, which I presented in *Hydrate the Earth*. The two scientists describe how the evapotranspiration of trees creates low pressure zones which draw in moist air to fill the void caused when liquid water rises, becoming a vapour and converting sensible heat into latent heat. When water vapour again condenses, becoming clouds and rain, another pressure void is created. This is because 1,000 litres of water vapour will condense into a single litre of liquid water. Gorshkov and Makarieva call this the biotic pump. Forests recycle moisture continuously, and in a tropical forest the process is daily, with rain falling every afternoon.

We saw that with Mediterranean summer storms, a critical percentage of water vapour was needed in the air for it to rain. The condensation nuclei emitted by vegetation are needed to trigger rain inland. Without forests connecting the coastal input of water vapour to the land, moist air will heat and dry as it moves inland. The process of moisture traveling inland and being cycled by forest corridors has been called "moisture hopping". Forests keep the moisture cycling and actually pull air in from the sea though the biotic pump.

Peter Bunyard, ecologist and author, says that 75% of the rain falling over the Amazon is from evapotranspiration, and 70% of the solar energy received in the area is converted to latent heat. That is a lot of cooling. The South American continent is bordered on the Pacific side by the Andes Mountains, so that moist air coming from the Pacific will fall on the western side of the mountains. The moisture coming from the Atlantic side travels 3,000 kilometres across the continent to the eastern slopes of the Andes. Because of the tremendous volume of water circulated by the rainforest, it rains more than it does on the coast.

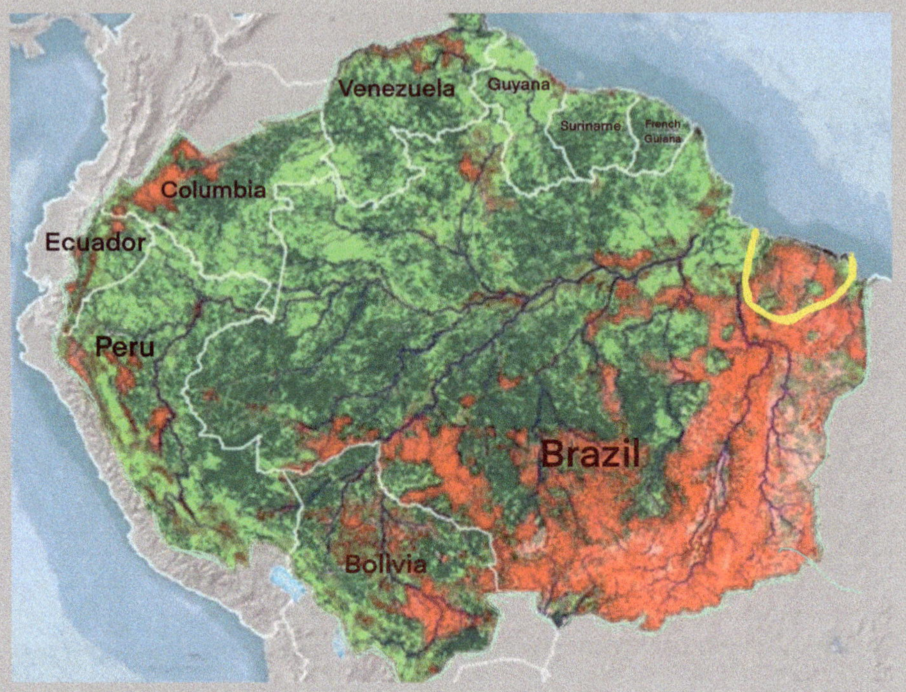

The Amazon Basin includes nine countries in South America, Brazil being the largest. The red areas show severely deforested regions, the light green less deforested regions, and the dark green areas that are intact.

The area outlined in yellow shows a critical coastal area in need of restoration to preserve the action of the biotic pump and the movement of moisture from the Atlantic across the Amazon.

Annual rainfall over the Amazon averages 2,500 mm. If the rainforest was to disappear, no rain would fall over the centre of the South American continent. The circulation of moisture is not limited to South America; it reaches the central plains of North America, down through Argentina, all the way to South Africa. Without it, the land will dry out, depriving millions of food and water.

Rob de Laet: Preserving the Biotic pump

Rob de Laet was born in the Netherlands, but has led a nomadic life and over the years has become a dedicated climate activist. In 2002, he ran an ecotourism venture company, bringing tourists on guided expeditions to natural sites around the world. He received an invitation from the Brazilian government to expand his operations there. Rob fell in love with Brazil and the rainforest. In 2007 he bought an apartment in the state capital of Bahia, Salvador, so that he could spend more time in the country.

In 2010, he became a climate activist, concerned that no substantial action was being taken to avert the climate crisis. He sold his company and bought a 250 hectares piece of land with a run-down house about 260 km inland from Salvador. The land had been deforested, converted into pastureland and subsequently degraded by overgrazing.

This is a common story in Brazil. Much of the land is either federally owned or reserved territory for indigenous peoples. Yet the land is often illegally occupied and cleared. On federal land, an occupant can claim ownership simply by virtue of having demarcated and cultivated an area. Once the land is degraded by overgrazing, it can be sold to grow soybeans, which will require chemical inputs to compensate for poor fertility. On the indigenous land this is illegal but the government does nothing about it. The land that Rob bought had been overgrazed, but he had no intention of growing soybeans. Instead, he was going to rewild it, restore the ecology.

Between 2012 and 2016, with the help of his mostly indigenous local friends, Rob planted 11,000 native trees, dug holes to collect and store rainwater and built a house fit to withstand the range of extreme weather, including rainfalls of up to 25 mm in 24 hours and prolonged, punishing heatwaves. The landscape restoration project was so successful that the average temperature in their valley decreased by five degrees and the house was sometimes too cool.

But Rob says that the biggest shift was personal. He had gone to restore the landscape and even envisioned establishing a community of like-minded people living autonomously. **And though he had always been a lover of nature, it was only here that he experienced, as he said, "the larger intelligence of the universe, an amazing flow of understanding."**

Another defining moment for Rob came in 2019. He met the brother of Antonio Nobre, Carlos Nobre, an esteemed climate scientist who has been studying the Amazon since the 90's. Thirty years ago, Carlos Nobre warned that with accelerating deforestation and advancing climate change, the Amazon rainforest could lose so much moisture that the forest would suffer dieback, becoming a savanna and triggering mass extinctions. Under the government of Jair Bolsonaro from 2019-2022, deforestation in the Amazon increased by 75% over the previous decade, destroying nearly 25% of forest cover that scientists predicted would push the forest beyond recovery. The Bolsonaro government dismantled environmental protection laws in favour of industrial agriculture, mining and logging. It turned a blind eye to illegal infiltration of protected areas, attacks against and even murder of environmental activists. Carlos Nobre became a well-known figure outside of scientific circles with the massive forest fires that raged in the Amazon during the dry season in 2019.

That year, he attended a conference in which Charles Eisenstein, author, environmentalist and public speaker, called the Amazon the heart, lungs, liver and kidneys of the entire planetary ecosystem. He said this ecosystem was heading towards organ failure. Rob then dreamt that the Amazon had died, and that a massive gathering of indigenous peoples from around world was mourning its loss. In his dream young people began to stream in from everywhere, artists, activists from every race and together they began to regenerate the planet. Rob was galvanised by this dream. Since then, he has devoted himself entirely to raising awareness and organizing ecosystem restoration projects for the rainforest.

Antonio Nobre pointed out that it is crucial to ensure the transport of moisture from the Atlantic Ocean to the forest. If instead, the land bordering the ocean is arid, the cooler air from the sea will draw the dry air from the land back to sea. But if there is forest, that mass of air mass will flow in the opposite direction, carrying moisture from the sea to the land. For this to occur, a closed canopy at least 100 km wide is necessary. A vast area called Arariboia lies just inland from the Atlantic coast. The territory is home to several indigenous peoples, most notably the Guajajara. A terrible forest fire devastated that area in 2015. Since then, some 10,000 square kilometres of forest has been razed for farming, ranching and mining. Rob befriended Ronilson Guajajara, a young chief of the Guajajara, who sought his help to reforest the territory. The Guajajara are good stewards of their forest; they know how to keep it healthy but not how to restore degraded land or stop the depredations of intruders. Since 2012, 28 forest guardians have lost their lives defending it.

An ambitious project was undertaken to enhance the transport of water from the Atlantic coast into the rainforest by planting trees. Rob set up a small NGO called Foundation for the Future of the Amazon. After developing a plan with the Guajajara, he sought financing for a three-tiered programme to regenerate a "Great Green Wall" of healthy forest to preserve the atmospheric moisture of the biotic pump.

The first stage has already begun. The goal is to plant around 16 million trees on 20,000 square hectares of severely degraded land. The Eden Projects Brazil NGO manages it, using funding to pay local indigenous people to plant the trees, and training workers in project management. It allows them to make the transition from traditional slash and burn agriculture to sustainable agroforestry. It was initially supported by charitable funding but the plan now is to increase that through carbon market financing.

The next stage will be to expand activities to include a newly developed approach called Assisted Natural Regeneration. The method is cheaper than replanting, it is based on removing barriers to regeneration, such as soil degradation, invasion by weeds, fires and grazing. The project includes a protection programme to limit damage caused by forest fires and illegal exploitation of the forest.

The last phase will emerge from consultations with the local people. Its aim is to create a socio-economic basis for maintaining the forest and the well-being of the local people through the creation of profitable agroforestry food production systems. The seed funding for these businesses will come through programme funding and be led by young local project managers trained in restoration work.

There is renewed hope since the defeat of the Bolsonaro government and the return of Lula da Silva. In his first month as President, he authorised sweeping changes. During his previous term, he had implemented a number of environmental protection policies. His cabinet includes Brazil's first indigenous minister, Sonia Guajajara, who was appointed to lead the newly created Ministry of Indigenous Affairs. Ms Guajajra said that she was there to "indigenise politics and reforest minds". The new environmental minister, Marina Silva, grew up in the Amazon and won prizes for her work as an activist in forest protection. This new government, which won the

Right: Huge areas of the rainforest in the Arariboia area were destroyed by forest fires, unauthorised farming and logging, turning the once dense forest into grassland.

election over the far right by only a slim margin, has instituted a policy of zero deforestation and is promising to enforce the state's power to prevent intrusions into protected areas. The government has made it clear to farmers that if they want to expand their agricultural production, they must improve degraded land.

Although these new developments give cause for hope, it remains to be seen whether the Lula government can prevail over the resistance of the far right. Many Bolsonaro supporters still hold positions of power, limiting what can get through Congress. This is where the will and the support of the international community could swing the balance in favour of the rainforest. There is progress being made in the movement to guarantee the protection of 30% of nature by 2030 with funds being promised by richer nations to support developing countries to preserve their biodiversity. But progress is slow despite the promise of change. Wildfires continue to ravage the Amazon, with 2024 being one of the worst fire seasons yet, while legal

and illegal logging, mining and agricultural exploitation continue. If the world recognises the Amazon as the treasure of biodiversity, carbon and water that it is, perhaps the pressure to destroy it for the sake of commodities such as wood, soybeans and precious metals will dieback instead of the trees.

Pablo Friedlander: Milking the Clouds, Bridging Cultures

Pablo Friedlander works at the opposite end of the Amazon watershed from Rob de Laet. He works mainly in the headwaters, in the high altitudes of the Andes Mountains. Each segment of a watershed contributes to the integrity of the ecosystem's biotic pump. Headwaters are tiny streams that collect in the highlands and as they descend and converge to become rivers.

Pablo's home is in the high altitudes of Champaqui Mountain in Central Argentina. This is the ancestral homeland of his mother's people. He returned here 21 years ago to make it his home after extensive travel and education. Being a Mestizo, the son of an indigenous mother and an Argentinean/ European father, he grew up as part of two cultures. When he was young, he was exposed to the shamanic traditions of the indigenous people. But he also pursued a classical European education, first in Argentina and then in Europe, spending ten years "in the ivory tower" of academia. He specialised in ethnobotany and humanities, earning his PhD in natural philosophy. He became an expert on the convergence of Western thought and indigenous traditional knowledge in new ecological paradigms of science and philosophy.

In 2002, he returned to Argentina and made his way home to Champaqui Mountain. He became involved with the local people, trying to understand the ecological issues troubling the area. Champaqui is the highest mountain in Central Argentina, and is approximately 430 million years older than the Andes. Many of the streams there had dried up. Pablo and his colleagues noticed that the remaining patches of high mountain forest were often damp. He concluded that trees at that altitude were extracting water directly from the clouds. Water vapour requires biological molecules to cause condensation. There are native trees specifically adapted to the highest

          Right: Pablo on a planting expedition among his beloved Polylepis trees.

altitudes which have the ability to capture moisture from the mist. They store it in fertile soils where it is purified by microorganisms, and then released progressively into the basin. This realisation was the starting point of Pablo's many collaborations. The necessity to restore the high-altitude forests became apparent. They called it the Milking the Clouds project.

The genus Polylepis comprises more than twenty species of trees native to South America, which grow at altitudes of 1,500-5,000 metres. It is related to the rose family. It's thick, rough bark allows it to resist cold temperatures. Polylepis store large quantities of water drawn from the clouds, even in the dry season. This moisture feeds the growth of mosses and bushes, which also store water. These remote Polylepis forests have been devastated by deforestation, cattle grazing and forest fires, denuding mountain slopes and summits. The few patches that remain indicate that water is still present in the ground, securing the life of streams and creeks that flow year-round, despite 6-7 months of drought most years.

Once it became apparent that increasing dryness and the loss of the high forests were related, Pablo and his associates contacted biologists from the

university to study methods of protecting and restoring the Polylepis forests. Pablo and his team began collecting native seeds in nearby forests less than 20 km from the area to be restored, grew the seedlings in special tubes during the winter and, come summer, planted them on the peaks of the Provincial Hydrological Reserve Pampa de Achala and the National Park Quebrada del Condorito. The original Milking the Clouds project expanded into a network of Polylepis reforestation projects called Accion Serrana. This bioregional organisation is part of Accion Andina, an international initiative active in most Andean countries. All the major rivers in South America originate in these high-altitude Polylepis forests, so there was much work to be done to expand the project along the range of the Andes Mountains. Many teams and partnerships have been formed to scale up and further this intergenerational work.

Since 2006, Pablo has been involved with the Institute of Ecotechnics, the same group that initiated the Biosphere 2 project. Those people were influenced by the work of a Russian scientist, Vladimir Vernadsky, who in the 1920s put forth the concept of the biosphere, of Earth as a living, conscious entity. The Institute of Ecotechnics aims to reconcile technology with this

view of a living Earth, which has a higher intelligence. They are involved in both research/educational and regenerative projects. Pablo participated in the Institute's training courses, taking part in expeditions, and eventually became a fellow of the Institute. His time there has influenced his work on the Milking the Clouds and Acción Serrana projects, and the Institute's ideas formed the basis of the Foundation of Biospheric Activities, which he set up with his associates in the Cordoba mountains and colleagues from Buenos Aires. Their world view is nearer to an indigenous world view, in contrast to a mechanistic understanding of ecosystems typical of western science. The Foundation has sought to expand the model of the Milking the Clouds project, encouraging local groups to begin planting trees in their areas. On average, 5,000 trees have been planted per year. In order to achieve the magnitude of their goal, more needed to be done and the Foundation enlisted international support.

Contacts in Europe enabled Pablo to set up the Treeangle Foundation in the UK. The aim was to create an international alliance of similar projects in three different regions: Champaqui in Central Argentina (Andes & Chaco Biome, with the Foundation of Biospheric Activities), Matutu in Brazil (Atlantic Rainforest Biome, with Matutu Foundation) and Montserrat in Catalonia (Mediterranean Forest Biome, with the Institute of Holistic Research of Montserrat). The work carried out on the Polylepis forests in Argentina has been adapted to other indigenous mountain tree species in these biomes. The Foundation has extended their work to include education and the involvement of local communities to look after these forests. International funding has helped the initiative, which has moved beyond grassroots efforts to create a whole social movement around tree planting.

It partnered with Accion Andina, based in Peru, which had replanted Polylepis forests in the high Andes in six different countries, including Champaqui Mountain. Instead of 5,000 trees being planted per year, the number soared to 150,000 in 2023, with 225,000 trees being grown in nurseries to be planted in Accion Serrana. This was made possible through

Left: Champaqui Mountain is the highest peak in Central Argentina. Seen here is a small patch of the native Polylepis forest, but most of the peaks have been denuded.

a network of partnerships ranging from local working groups, international organisations and scientific collaborators. Local communities do the actual planting, but they are monitored and trained by experts from international organisations. Financial support pays for the training and the salaries of the mostly native population taking part in the initiative. Now that the project is scaling up, they have received proposals of further funding for scientific research. They are carefully screened to be sure that they align with the ethical values of the project. In terms of funding for carbon sequestration, high altitude forests have less potential than rapidly growing tropical forests, but they are valuable for the water, soils and biodiversity that a healthy ecosystem preserves.

Beyond the restoration of high-altitude forests, once ecosystems stabilise, they provide opportunities to develop businesses based on products supplied by them, providing further support to local communities. One such initiative is Champatea, a herbal tea blend made from yerba mate mixed with wild herbs which grow naturally around Polylepis forests. Pablo partnered with friends and an entrepreneur from the UK to start a company, fully owned by the Treeangle Foundation, that markets products from communities around Champaqui and related regenerative projects. Although the establishment of the forests depends on grants and donations, the hope is to one day sustain the work through revenue streams generated by the ecosystem and run the local communities around them.

Since the Milking the Clouds project and Accion Andina has expanded, Pablo recently accepted to work with colleagues who have set up similar project in Brazil. El Puente, which means The Bridge, is a charitable foundation helping indigenous communities in the Amazon Basin defend and restore their lands. It supports projects developed and led by these communities which have a tradition of working with "teaching" or "medicine" plants, a valuable source of knowledge for the development of new medicines. The native-run projects and enterprises also provide economic support for preserving the ecosystem that sustains them.

There are four pillars to the El Puente project. **The first is Sovereignty.** Above all, the indigenous territory and their sacred sites are to be secured and protected from illegal mining, logging and other intrusions. Communities govern themselves, making their own decisions about how their lands are managed. The aim is to secure lands for reserves and build infrastructure for each project.

The second pillar is Regeneration. The protocols for regenerating land around a river are very different from those required in the mountains. In some places planting indigenous tree species is required, in others assisted natural regeneration is sufficient. The main focus of this pillar is to establish Ethnobotanical Reserves, from which all the main food sources and medicines are derived.

The third one is Education. Local people need to be trained in restoration work. But education works in the other direction as well. Local communities possess ancestral knowledge concerning methods of agroforestry or harvesting and processing of natural forest products such as cacao, to name only one. Training in practical skills applies not only to local communities, but to collaborators, who learn the traditional methods and values of indigenous peoples. The many foods and medicines known to the indigenous communities could benefit both worlds if the relationship is built on mutual respect. El Puente offers a bridge between cultures by first recognising indigenous sovereignty and knowledge, then by educating both sides and finally building a relationship of "Reciprocity".

Reciprocity is the fourth pillar. It requires mutual trust and respect. The work on the first three pillars enables the fourth. When indigenous communities know that their ways and lands are being protected, and they have the power to conduct their enterprises as they see fit, only then is reciprocity possible.

The project is being carried out in four native villages along Amazonian rivers. As these sites develop, they will provide models for future projects in other biomes. They also bring Andean and Amazonian communities closer together. The overarching goal is to bridge the best of indigenous wisdom with the power of western technology to unite people with good intentions everywhere, to fund regenerative projects and educate people.

Pablo is a visionary, a poet and a philosopher at heart. His work is founded on a central view of the Biosphere as Pachamama, the mother of Earth, a living being, a cosmic phenomenon powered by the Sun, Tata Inti. His background in western science and technology provides him the means to connect and unite peoples around the world with shared knowledge and values.

Natural biomes

A visit to any international city today is proof enough that globalisation has eradicated many regional differences. We find the same fast-food chains, street signs and shops. A similar conformity has been imposed on the landscape. Industrial agriculture is everywhere, attempting to grow the same selection of food crops which make up the modern diet everywhere. But nature is not a conformist. It suffers under the practices imposed by a globalised culture. Such practices have decimated natural biomes which have evolved over millennia to thrive in the specific conditions of different parts of the world.

The flora and fauna of an ecosystem is uniquely adapted to the conditions there. A biome is larger than an ecosystem and may include several of them. The amount of rainfall, the number of hours of sunlight, temperatures, altitude and the type of soil are all critical factors determining what will thrive. There are both terrestrial and aquatic biomes. For our purposes here we will focus on terrestrial biomes.

The **tropical rainforest biome** is found near the equator and hosts the greatest biodiversity in the world. It is exposed to equal amounts of daylight year-round and is always warm. It also rains year-round, receiving between 2,000-10,000 mm annually, more than any other region. There are three different stories of dense vegetation: from tall trees to thick vines and an understory of bushes and various plants. Each story is home to a huge diversity of animals, birds, insects and microbes.

The **temperate rainforest biomes** lie further north and south of the equator, and are situated on continental coastlines. These forests are humid, the rains are constant, but they experience a drier spell in the summer months. Because they are coastal, temperatures are typically milder in summer and winter than they are further inland. They have only two layers of canopy: tall trees and the understory. The soil is rich in organic matter and densely covered with mosses and ferns.

Temperate deciduous forests are located in regions where temperatures vary greatly between winter and summer, as does the amount of light. These forests receive 750-1500 mm of rain throughout the year. There are four seasons, and trees are adapted to the extreme temperature shifts by losing their leaves in the autumn and regrowing them in the spring, which creates a rich sponge of organic matter colonised by mosses. These forests are home to a great diversity of animals, birds and insects, which are either migratory or have adapted to seasonal changes.

Further inland lie the **temperate and tropical grasslands or savanna**. These flat areas receive 500-900 mm of rain annually. The height of the grasses depends on the amount of rainfall. Tropical grasslands have a wet and a dry season, while temperate grasslands freeze in winter. Sod protects the roots of the grasses so that even if they die off in the winter or the dry season, they will regrow once the season ends. Streams are found in grassland biomes. Trees are few and small because of infrequent rain.

The **taiga biome or boreal forest** dominates landscapes below the Arctic circle. Winters are long and dry while summers are short and wet. Most of the precipitation is snowfall. Trees are mainly coniferous with an understory of low-growing plants. Animal species include bear, moose, hare, deer and woodpeckers. Much of northern Canada, northern Europe and Russia is taiga.

Beyond the taiga, in the highest latitudes, we come across the **tundra biome**. Vegetation receives about 60 days a year of sunlight and warmth. No trees grow in the tundra, only shrubs, lichens, mosses and sedge grass. It is home to caribou, lemmings, fish and migratory birds, along with plenty of flies and mosquitos.

The **alpine biome** exists in high-altitude mountain ranges. Trees cannot survive at these altitudes, but shrubs, grasses and heaths are prevalent. Sheep, elk, goats and little furry rodents called pikas happily dwell in the high peaks, along with birds and insects adapted to the thin air.

The **chaparral biome** receives about 630-750 mm of rainfall annually, mainly in the winter. The plants are woody shrubs which go dormant in the dry summer months. They are generally evergreens with thick waxy leaves which retain water. Chaparral is often found in coastal areas such as California, Mexico, the Mediterranean Basin, Chile, South Africa and parts of Australia. The animals of the chaparral biomes include coyotes, lynx, pumas, packrats, sheep, boar and rabbits.

Less than 400 mm of annual rainfall falls on the **desert biome**, and sometimes it does not rain for years. When it does rain, it is usually heavy and brief. Animals and vegetation are adapted to survive in extreme temperatures with little water. Plants are succulents like cactus or yucca. Animals include rodents, owls, lizards. They protect themselves by burrowing into the ground. Deserts are generally found inland north and south of the equator where the Hadley cell meets the Ferrel cell, which results in drier air with high pressure because the moisture absorbed at the equator has been spent.

Many parts of the world are in transitional zones where different biomes meet and overlap, so are less easy to characterise.

It is important to recognise that the natural ecosystem which is adapted to its unique biome has evolved over millenia to be self-sustaining. The interactions of the flora and fauna have feedback loops which regulate and sustain life. When we alter these systems by removing too many of the key players, we unwittingly cause breakdowns in the ecological balance.

Ivan de Klee: Rewilding the land, valuing natural capital

Ivan de Klee is a leader in the growing movement to revive nature in the UK. He is helping to build the infrastructure to "rewild" a significant percentage of the land, in order to restore biodiversity and the ecosystem services essential to our survival.

When Ivan finished school, he had no clear idea what he wanted to do. He loved nature and the great outdoors but didn't really see how that could be a career path for him. He enrolled in an undergraduate programme in World Religions. In his fourth year, he registered for a course in Ecology and Ethics, which examined big issues such as global warming, overpopulation and the threats to biodiversity. This exploration called to him and he knew he had to be a part of the solution. During this time, he read Michael McCarthy's book, *Say Goodbye to the Cuckoo: Migratory Birds and the Impending Ecological Catastrophe*. It discusses the alarming decrease in the population of migratory birds, due mainly to the loss of habitat. The population of cuckoos, the beloved harbinger of spring in the British Isles, has dwindled to 37% of what it was in 1967; populations of other migratory birds were also in decline. Ivan felt called to career in nature conservation.

Although he had a degree in World Religions rather than a field related to his ambition, he didn't't want to return immediately to university. Instead, he volunteered for nature conservation projects, at first in the UK and then in Africa. He worked at a rhinoceros reserve in Kenya, then in India with Belinda Wright, a well-known conservationist dedicated to the protection of wild tigers. It was there that he heard about a radical conservation project back in his native England, called Knepp.

Knepp Castle Estate is a 1,400 hectares property in Sussex, England, which transitioned from a conventional dairy farm to a rewilding project called Knepp Wildland. When Charlie Burrell inherited the Knepp Estate from his grandparents in 1987, it was an unprofitable dairy farm. Prior to the 13th Century, Knepp had been a wooded hunting ground inhabited by fallow deer and wild boar. During the second World War, the British government encouraged the clearing of land as part of a "Dig for Victory" programme to provide food self-sufficiency in Britain. The soil at Knepp is a heavy clay soil, completely unsuitable for growing crops. Although Knepp was a dairy farm, it only survived thanks to farm subsidies.

Charlie and his wife, Isabella Tree, tried valiantly to make the farm profitable right up until 1999, at which point they decided to sell their herds and equipment and abandon farming. But in 2000, they came across a copy of Frans Vera's book, *Grazing Ecology and Forest History*. She challenged the commonly held notion that before the arrival of homo sapiens, the British Isles and northern Europe were closed canopy forests. Vera claims that the landscape was in fact shaped by wildlife. Animals are disruptors, fighting against the tendency of plants to evolve and disseminate until they reach a "climax ecosystem", or a fully mature, closed canopy forest. Animals constantly graze and nibble the plants, preventing them from taking over completely. Every species of animal acts on the environment in a different manner, and the resultant ecosystem arises from the balance struck between natural plant succession and the wildlife that limits its spread. The richest

biodiversity occurs at the edge of two systems. The tension between shade dwelling species of plants in a forest with the Sun loving species in an open field is further enhanced by the activities of animals in these systems.

Vera himself founded a 6,000 hectares conservation site in the Netherlands called Oostvaardersplassen. The site had once been a nursery for willow trees, but to test his theories, Vera had populated it with large herbivores in an attempt to recreate what might have been the original biome. In Europe, before they were driven to extinction, there once roamed wild ponies called tarpan and wild cattle called aurochs, along with deer and boar. Vera's idea was to release functional equivalents of these extinct herbivores and study their impact on the landscape. The animals were not tended but simply left to fend for themselves year-round.

Charlie and Isabella knew after visiting Frans Vera and Oostvaardersplassen that they wanted to try wilding at Knepp. The first fallow deer were introduced in 2002. That same year Charlie sent a letter of intent to the Department of Environment, Food and Rural Affairs informing them of his plan to establish a biodiverse wilderness area on his land. In 2004, twenty-three English longhorn cattle were released, as well as two Tamworth sows with eight piglets and an Exmoor pony. The land was left fallow. The experiment was underway to see how these herbivores would shape the landscape. For practical reasons, the land was fenced to prevent the animals from wandering onto the highway. Nor was it possible to introduce natural predators. Instead, the animal population would be culled through harvesting wild meat. The sale of meat would also provide income to the farm.

Over the years, native and endangered species have gradually returned to the landscape. Residents include turtle doves, nightingales, bats, owls, the rare purple emperor butterfly, native bees, and peregrine falcons. The landscape continues to evolve as it becomes wilder, readjusting its balance as the ecology becomes more complex and diverse. In 2014, Knepp Wildland, campsite and safari business opened, offering tourists a chance to witness the splendours of the wild through guided tours and "glamping": charming and comfortable campsites which are rented out. Knepp is now a haven for a host of rare and endangered species where people can come and enjoy a wild natural space.

The British Isles are accustomed to rains and floods because of their location at the meeting point of the Polar and the Ferrel cells, which create a low-pressure zone. But climate change has exacerbated the potential for extreme flooding. In the Victorian era, a significant portion of the British landscape was altered to control the flow of water. Rivers were artificially deepened; canals were dug to transport goods. Wetlands were drained and filled to allow for farming and housing. As part of their rewilding of Knepp, Charlie and Isabella were inspired by the success of a "Room for the River" programme, which had been successful at mitigating floods in the Netherlands. They started to restore the floodplain in a section of the Adur River that flows through the Knepp estate. They removed the banks of the canal, which allowed floodwaters caused by heavy rain to overflow onto the plain, creating gullies and marshes. Whereas the banks of the river beyond Knepp are steep and devoid of wildlife, those at Knepp are home to marshland birds, insects and invertebrates. The restored river filters out sediments and farmland pollutants from the water flowing through Knepp,

and the extra water absorbed by the landscape serves to control flooding downstream of the property.

Charlie and Isabella became aware that a key species was missing from Knepp: the beaver. The Eurasian beaver contributed to shaping the hydrology in Europe. It became extinct through aggressive hunting and loss of habitat. Much of the work done at great cost at Knepp with heavy machinery and work crews would have been done for free by beavers. Charlie founded BACE (Beaver Advisory Committee for England) in 2010 as a partnership of conservation groups advocating for their reintroduction. It was not until January 2020 that they were granted permission to reintroduce beavers at Knepp. Now two beaver pairs have made their home there and will continue to work the hydrology of the area to assure that plenty of water is stored in the ground, ensuring resilience to droughts and floods.

In 2015, after hearing about the Knepp rewilding project, Ivan de Klee wrote to Knepp asking to work on a project while he was on leave from the safari lodge in India during monsoon season. He was hired as a farmhand, and spent six months repairing fences, tending the cattle and doing some biological monitoring. Yet Ivan was still not working as a conservationist. He signed up to do a Master's degree in conservation science, but at the moment he was to return to university, he was offered a position working for Flora + Fauna International in South Sudan. He accepted the post, his first bona fide, properly paid conservation job before returning to the UK to complete his Masters. Once he had the degree behind him and he was qualified as a conservationist, he returned to Knepp and became a part of the team there.

Compared to the often frustrating experience in conservation in South Sudan, he found the focus on regeneration at Knepp refreshingly positive. Instead of a desperate struggle in a losing battle, rewilding gave Ivan hope that he could help bring about a better future. Far from the traditional conservation work based on management and target setting, rewilding focusses on minimal intervention, simply by restoring key species to the

Left: Within an ecosystem, the landscape is shaped by the tension between plant species and animals. The oddly shaped vegetation at Knepp Wildlands were made by browsing free range animals.

At Knepp, English Longhorn cattle are the functional equivalent to the extinct aurochs, wild ruminants which once roamed the countryside. They keep the forest canopy open for light loving trees such as native oak.

landscape and removing human obstacles. There was no doubt that this was the work he wanted to do.

By the time Ivan arrived at Knepp, Isabella had published her 2018 book *Wilding: The return of Nature to a British Farm.* The natural progression of the landscape and the supporting businesses were well under way. Charlie and Isabella were turning their attention to promoting the rewilding model. Ivan took part in this initiative. Although the Knepp Estate is large, it is still only a small patch of wild in an expanse of conventionally managed land. Ivan's first project at Knepp was to lead a group of about twenty-five farms from the surrounding area in an exploration of landscape regeneration concepts. He organised workshops on various aspects of regenerative farming, from grass fed animals, to creating habitat niches on farms to encourage biodiversity. His aim was not to persuade farmers to rewild, as Knepp had done, but to modify their methods in order to extend the natural habitat.

Right: Charlie and Isabella restored the river banks to their natural state, converting steep level channels into natural slopes so that the river can overflow into the floodplain. The earth absorbs the excess water, replenishing the aquifers.

He also helped farmers access grants and subsidies to make the changes they were interested in making.

This farm cluster exploration developed into a broader project called Weald to Waves, which envisioned the creation of a connective habitat for wildlife extending from the Ashdown Forest in Sussex (home of Winnie the Pooh), where Knepp is located, down to the coast. The project aims to help farms along the watershed become more nature friendly through rewilding, regenerative agriculture, sensitive woodland management or managed conservation. The goal of the project is to defragment the natural landscape, making a corridor for nature to circulate. Creating connectivity of natural spaces helps us to coexist with nature. It is better for nature to have a connected flow than to have tiny disconnected patches of habitat scattered through the landscape.

The cost of the wild

The need for funding has become the latest focus at Knepp. Ivan's role has now transitioned to a new project called Nattergal, aimed at creating investment products to fund rewilding. The World Economic Forum estimated that to reverse biodiversity loss by 2030, we need to increase

funding for nature by an average of 711 billion USD annually. This figure is based only on what it will cost to stop the degradation; regeneration will cost even more. The funding gap will not be met by government subsidies alone. The bulk of these financial flows will need to come from corporations and wealthy investors. The goal of Nattergal is to bring large flows of money into supporting ecosystem services: carbon sequestration, clean water and biodiversity, which at present are given away for free and are therefore undervalued. The World Economic Forum reported in 2019 that the fate of the global economy is dependent on the fate of nature and that global industrial culture is destroying nature faster than it can renew itself.

Nattergal was founded in 2021 by Charlie Burrell along with Jeremy Legett, founder of Highlands Rewilding and former Greenpeace executive, and Pete Davies, a fund manager and bird enthusiast. The founders are non-executive Board members, and there is a team of seven people working on the project. Two sites have been acquired and are being monitored to establish a baseline of the state of ecosystem services. One of these sites has a permit to house the largest so far beaver enclosure in the UK. Investors in the purchased land receive a return based on the augmentation of ecosystem services. With the exception of the carbon market, investment in ecosystem services is still small. Nattergal is hoping to expand it.

There are programmes coming into effect in the UK, Australia and elsewhere that will require entities such as housing contractors, whose projects have an impact on natural environment, to compensate financially for the loss of biodiversity plus 10%. The level of biodiversity loss cannot be calculated precisely, but will be estimated according to the quality of the environment. For example, a natural forest will cost more to offset than a field of wheat. There is a budding voluntary biodiversity market, which was set in motion following the December 2022 Biodiversity COP in Montreal, where 190 nations signed on to the Global Biodiversity Framework, an ambitious set of targets to halt biodiversity loss. The UN has launched the Task Force on Nature-Related Financial Disclosures (TNFD) which is putting a framework into place to set up best practice standards and baselines for corporates to assess their risks and disclose their behaviors. All of this will certainly help to evolve the financial mechanisms and incentives to drive the market for biodiversity investments.

The market for water credits is driven by a concept called nutrient neutrality. Natural England, an independent body created by the UK government, is working on water protection guidelines because of concerns about high levels of pollution from farm chemicals, stormwater and wastewater. Before being granted a construction permit, developers must now prove that a project will not add to the nutrient load in a water catchment. Procedures are being crafted to calculate the nutrient load within a water catchment area and an offsetting programme will soon come into effect, similar to the carbon market. Financing for future wetland rehabilitation may come from water companies or development projects seeking nutrient neutrality by offsetting their impact. With polluters obliged to take responsibility for their messes, there will be more incentives to build green infrastructures for water purification. In the long run, it may be less costly for developers to invest in green infrastructure than to continue offsetting. Offsetting will also make funds available for restoring wetlands and water retention landscapes.

As this shapes up, Nattergal will be well placed with their scientific monitoring and the considerable experience gained at Knepp. Ivan is well aware of the dangers of commoditising nature. In an ideal world, he says, we would value nature intrinsically and have a more spiritual attitude towards life, but given the world economy runs on a system that accounts for value, we cannot accord nature a value of 0.

We are in a state of emergency, and creating a market for ecosystem services is our best choice in the short term for financing effective action.

Chapter Four
Regenerate the land, regenerate the culture

Indigenous Culture and Traditional Ecological Knowledge

At one time, all people were indigenous. The word can be applied to any species; it means "native to the land". Indigenous species are those that have adapted to a particular topography and climate. Modern society, however, poses a threat to these cultures. Modernisation, wherever it has occurred, has displaced or absorbed traditional cultures. Those that remain are struggling to maintain their way of life. The World Economic Forum reported that lands still managed by indigenous people account for 80% of the world's biodiversity, despite the fact that they represent only 5% of the global population. Modern society has for too long dismissed these peoples as primitive. Their way of life appears simple, but it is informed by thousands of years of close observation of natural processes. Modern science is just beginning to recognize that these peoples, who were once considered savages actually possess a wealth of what is now being called traditional ecological knowledge. In this time of planetary crisis, scientists have begun to value the depth of that knowledge and seek it out in order to work together with indigenous peoples to find solutions.

Globalisation has surpassed the limits of growth. In 2009, a team of Earth system and environmental scientists under the auspices of the Stockholm

Resilience Centre and Australian National University, determined nine 'planetary boundaries' necessary to regulate and stabilise our planet. In six out of nine of these planetary boundaries, we have exceeded the limits. If all humans on Earth lived the lifestyle of the average person in a wealthy developed country, it would require five Earths to sustain us.

Many indigenous peoples survive today, and each one is unique. What may safely be generalised is that indigenous cultures have been far more successful than our globalised society in preserving nature. What differentiates indigenous cultures is that their ethics are based on the belief that human beings are part of an interconnected web of life, and are therefore responsible for its care. This conviction is not merely private, but collectively shared through ceremonies, rituals and stories, the foundations of their communities.

Modern science has just begun to discover that nature is complex and interconnected. We now know that the idea of the biosphere as an interconnected, intelligent system is a biological reality. But these recent discoveries are not yet integrated in our culture, which does not support an ethic of care for the Earth. Much as we, as individuals, may value nature and relationships, we are all in the grips of a system which does not incentivise protecting what is of true value. We all need clean water, fresh air, healthy food, shelter and community to thrive. How can we create a global culture which gives importance to what is vital for our survival?

If we were to investigate those cultures that have been more successful than we in preserving the pillars of life, would we remember our roots? If so, it might transform our wasteful consumer culture into a regenerative one and reawaken the sense of kinship with all the living beings.

Create a water ceremony

On the night of the full moon choose a cup or glass and fill it with potable water. Place it either outside in a special spot or — if is too cold outside — on a window ledge inside. Go outside and look at the moon, and while doing so, think about water, how your body is made of water, how we need water to live. Express your gratitude for water to the moon. Tell water you love and respect her. Maybe sing a song for water. The next day, drink the cup that spent the night in the moonlight and feel gratitude for water again.

Afterwards, have a discussion with the group about how everyone felt to treat water and the moon like that? Did it feel silly? Did you feel grateful? What was it like to go outside at night to see the moon?

Talk about how indigenous people relate to the natural world as beings and relatives while we relate to it as commodities for us to use.

If you actually did the above exercise, you may have felt silly and awkward. If you went outside to look at the full moon you might have been surprised to feel the night air and look at the sky. Most of us have lost the habit of going outdoors and simply being in nature. We are caught in our to-do list, sitting in front of a screen. Maybe we need to find new ways to create a culture of connection. Not necessarily adopting anyone else's way, but being open to other ways of being and thinking.

The Mi'kmag are the First Nation inhabitants of Canada's Atlantic Provinces. Mi'kmaq Elder, Dr Albert Marshall, teaches a concept called Etuaptmumk, which means "Two Eyed Seeing". Just as we need both our eyes, each seeing from a slightly different perspective, to perceive depth, we need to reconcile Western and indigenous ways of knowledge. The elders have used the metaphor of different species of trees "holding hands" with their roots. The concept of Two Eyed Seeing was developed to guide reconciliation between the global culture and the local indigenous people, acknowledging that there is merit in both cultures.

We already have the know-how to restore ecosystems. Within a short period of time, restoration brings benefits: in ten years we can rehydrate a landscape, rebuild soil health, and revivify native species. Nature knows how to heal if we cooperate with her.

There are many projects that exist as a patchwork of efforts around the globe. In many countries, in widely differing contexts, the principles introduced in this book are being applied and perfected. The next stage is to scale up, especially in this decade, when the future of our civilisation is being decided. For this, we need a more global shift of culture and mind set.

Marius Iragi Ziganira and Owen Allen: Building a regenerative culture in a refugee camp

Marius Iragi in Africa, Owen Allen in Australia, came together by chance and created hope in a very difficult situation. Their story shows how international partnerships facilitated by modern technology can transform the world.

Wars and climate disasters result in an increasing number of refugees around the world. We can expect that as climate change worsens there will only be more displacement. Too often, refugees find themselves in camps, where they remain for an indeterminate amount of time, living in degrading conditions and dependant on food aid. Many refugee camps suffer extreme overcrowding. Hygiene and sanitation are poor; people are forced to clear vegetation to make shelters and they may overburden water resources. It is a situation that degrades both the land and the spirit.

Right: Marius Iragi Ziganira in one of the regenerative gardens of Nakivale refugee Camp in Uganda

The Nakivale Refugee Settlement was set up in southern Uganda in 1958. It covers an extraordinary 185 square kilometres, and is home to some 130,000 refugees, many of whom have lived here for decades. They live among the native Ugandans who were in the area before the camp was established. Small villages are scattered throughout the territory.

Marius Iragi Ziganira was born in the Democratic Republic of Congo (DRC) in the town of Bukavu. He pursued his education up until university, where he studied information technology and e-commerce, then switched to agronomy, specialising in soil science and environment. After his studies he worked with Action Sociale et Organisation Paysanne (ASOP) as a supervisor of the Agriculture Adapted to Climate Change project financed by Oxfam. He left the position to start a chicken farming operation in his home village with the aim of improving food security, reducing unemployment and educating young people in sustainable development. There he built a facility which housed 2,000 chickens. In addition to raising and selling chickens, he took on interns, training them in breeding techniques and

small-scale production of eggs, and providing them with hens to start their own productions. Between 2014 and 2018, he mentored 150 young people.

In 2018, his life was devastated when rebel forces attacked his village. DRC is in a constant state of civil unrest. Warring factions and massacres are not uncommon. He lost his farm. The village was destroyed, women were raped; Marius's wife was raped in front of him. People fled. Marius and a small group of villagers made it to Goma, the provincial capital, where some priests helped the group make it over the border with Uganda and to the Nakivale Refugee Settlement. He was separated from some members of his family, and to this day does not know what happened to them.

Food aid came from international organisations. At the end of every month, each person received an allocation of rice, beans, maize and oil. They were allotted a small plot of land to grow food, but most had no experience of farming. After a time, food rations were halted and replaced by a monthly stipend, currently about USD 3.60 to purchase food. The stipend was insufficient, and some left the settlement in search of other options. Marius tried to teach people the skills he practiced in his home village. Many of the refugees who benefitted were single mothers interested in becoming farmers.

During the COVID19 pandemic, people in confinement stayed in touch with one another through internet as never before. Marius was no exception. He had always been concerned about climate change and environment, and he began to attend virtual meetings of the Foundation for Climate Restoration (F4CR). It was through these online meetings that Marius and Owen became friends and began the fascinating virtual relationship that has transformed their lives.

Owen lives in Atherton, North Queensland in Australia. In 2019, he was in a process of re-evaluating his life and his career. He was a physiotherapist nearing 60 and wondering how he might spend the rest of his productive years doing the most he could for humanity while requiring the least resources for his own survival. In 2019, Australia suffered devastating bushfires which

Right: Owen Allen on Phoenix Farm in his native Australia. Owen has a vision of growing tropical fruit trees and vegetables here, a delightful, productive environment managed by a community of growers, artists, ecologists.

destroyed vast swaths of land and cost the lives of millions of animals. Having just closed his practice, Owen set out on the road to help with bushfire relief. But before he could get there, authorities closed the borders because of the pandemic. Back at home he joined Zoom meetings with international groups, including the F4CR, which was focussed on initiatives to lower carbon dioxide to pre-industrial levels.

Owen was paired up with Marius in an effort to explore how the international community could help advance carbon sequestration efforts in Africa. After several group meetings, Owen and Marius began to have almost weekly meetings on Zoom, initially around a seminar that Marius was planning for International Women's Day. Throughout 2021 they continued to meet to assess the needs of the community and to plan more seminars, inviting people from all over the world to give conferences and presentations virtually to groups of the Nakivale refugees. The relationship that developed between Marius and Owen is particularly remarkable because, as Owen speaks only English and Marius speaks both Swahili and French, they communicated through a translator. The two men have developed a strong bond over the past several years, yet they have never met in person.

Owen organised a fundraiser on GoFundMe for a regenerative agriculture project. By January 2022, they had collected AUD 490. They were used to purchase maize and bean seeds for 18 farmers, who would grow their own crops, save enough seeds for the next planting season, donate the rest to a seed bank, which would supply other farmers as the project expands Over the course of 2022, around 65 farmers were provided with seed. It looked as if the programme was running to expectations, until a freak hailstorm ruined a major crop during flowering. The poor return of that crop not only dismayed the farmers but, in failing to provide a return, forced many to travel outside of Nakivale to find work.

The set back forced Marius and Owen to revise their plans. and they chose a 'demonstration' farm model so that they could have more control of the agricultural outcomes and provide lessons for local farmers. The demonstration gardens will also continue to provide returns to the seed

bank and the farmers gain both the seeds and the skills to continue their home gardens.

In addition to the regenerative agriculture project, Marius and Owen turned their attention to other needs in the community. There were many single women and lots of children, people with disabilities, including albinism, a genetic disorder characterised by a lack of pigment in the skin; individuals in Africa who suffer from it are often marginalised because of superstition. Marius organised a team to carry out a survey of some 175 people with disabilities in the Base camp zone of Nakivale. In 2022, working with the Heidi Latsky Dance Company project, On Display Global, Owen provided the world with a glimpse of the people with disabilities in the camp. It is hoped this will be an annual contribution for awareness raising. To meet the occupational needs of disabled people in Nakivale, Marius and Owen are considering the construction of a cultural and skills training centre.

The refugee settlement is overseen by the Ugandan government and the United Nations High Commission for Refugees (UNHCR). They occasionally consult local stakeholders and evaluate proposals for development projects in the area. A local officer at the Ugandan Department of the Environment, who was familiar with Marius' and Owen's seminars, asked them to propose a project at the beginning of 2022. The Office of the Prime Minister and UNHCR were impressed by their proposal, and requested that Marius and Owen upgrade the organisational structure of their programme by creating a foundation. A new organisation called the Umoja Community Foundation (UCF) was registered. "Umoja" means "Strength in Unity". Marius and Owen serve respectively as the CEO and the Chairman of the Board, whose five members oversee the work of the foundation. Although international representatives are allowed to sit on the board, currently everyone except Owen resides in the Nakivale settlement.

The Ugandan government also donated chickens to the project, both for food and to provide manure for fertiliser. Young people and those with disabilities are being trained to care for the chickens. The regenerative

Left: Marius oversees the distribution of seeds to the first cohort of farmers in the regenerative agriculture program.

agriculture project is growing despite challenges. Inexperience and unpredictable climate frustrates the efforts of many farmers, but help from international organisations to train them in eco-agriculture and new farming techniques encourages them to continue.

Three international specialists in carbon sequestration and carbon markets have partnered to donate biochar kilns to the UCF as a proof-of-concept project on carbon sequestration. Biochar is a soil amendment made from burning woody waste material in a special kiln with low oxygen until it creates a kind of charcoal-like particle which sequesters carbon. It can hold the carbon for over 100 years and stimulates the growth of soil microorganisms. With these kilns, the community will be able to burn waste materials from sawmills, farming operations, and community waste to create biochar. The kilns have special sensors which can send data through a mobile phone to a central location for monitoring. This helps international partners collect data remotely for their research. The biochar will be used to improve local soils and the partner groups will consolidate data on carbon sequestration so that eventually they will be able to claim carbon credits as additional revenue.

Upon registration, the UCF opened a bank account to permit microfinancing. As a first step, it offered a savings service to women who were market stall holders, allowing them to save for family needs such as purchasing clothes and educational material for their children. Today this microfinance programme serves more than 190 members, mainly women, who carry out small income generating activities.

Owen's coincidental connection with African Leadership Transformation Foundation (ALT), an African organisation committed to radically transforming Africa through leadership training, provided an opportunity for refugee women to access a leadership course, which also dealt with mental health. When Owen met with Daniel Kamanga, the founder of ALT, to discuss offering courses at the refugee settlement, he thought it was a brilliant idea; as he had never thought of reaching out to refugees, which number some 30 million in Africa. As of March 2023, 85 Nakivale refugee women had taken the course. It is expected that another 30 women will enter the training programme in 2024.

Owen invited Katrina Mellick, a career occupational skills trainer and employment agency manager in Australia, to design a youth leadership programme. It began in January 2023 with 30 participants, who attended 30 modules, each focussing on a different aspect, after which they were divided into small groups to design a project. This "study circle" concept is expected to be effective in creating a learning culture in Nakivale.

Both the women and youth leadership programmes train a cohort of the trainees to become translators and facilitators. Multiplying the impact of programmes by "training the trainers" is one of the guiding principles of Marius' and Owen's work. The team now has an international group of experts in eco-agriculture and ecosystem restoration committed to the Nakivale settlement. Work with these internationals is now becoming strategic around a 20-year future, and UCF expects to see increasing progress on an annual basis, including joining the global food and business market.

Marius is sometimes overwhelmed by the need and the suffering he sees around him, but rather than give in to despair, he has taken action. There is so much more work to do and the road ahead is long. He believes that the model he and Owen developed at Nakivale can benefit other refugee camps. He is looking to form partnerships with similar organisations around the world, who share his aim to train refugees to develop sustainable food and economic activities, manage the resources of the surrounding ecosystems, and build a regenerative culture.

Maya Dutta: Bringing nature into the city

Maya Dutta lives in Cambridge, Massachusetts in the United States and she has found her passion planting forests in the city, using a method known as Miyawaki, which is the creation of tiny niches of native habitat. When I asked her to summarize her project, she said: "We plant small, biodiverse, densely planted native forests that create the forest ecosystem natural to an area. It can be done in very small areas, about a thousand square feet is the minimum size. In just six parking spaces in the corner of a lot, in the front of a school yard, you can transform bare ground into a habitat, a place that cools and beautifies. It can bring nature back into people's lives, they can access the chance to be close to and steward an ecosystem. It's so important for general well-being."

Although Maya's parents are from India, she was born in the 90's in New York City. She became aware of climate change and suffered from eco-anxiety at a young age. When she was ten, after watching Al Gore's film *An Inconvenient Truth*, she was frightened that there would be no future for her. As she grew older, these anxieties disappeared into the background. She went to university, studying maths and science, and found herself a steady job in the tech industry. However, her anxieties about climate change continued to trouble her.

One day while visiting a botanical garden, she felt a deep sense of connection to the plants and trees, followed by a sense of heartbreak about the loss of nature. Although she felt the calling, a part of her resisted. A few years later, during an Earth Day screening of an animated film, *Princess Mononoke*, a story about a struggle between forest spirits and a mining company, Maya sobbed throughout the film and all the way home. The film reawakened her conviction that she needed to be a part of the solution. After considering many options, such as returning to university for training in climate modelling, she came across ideas about regeneration. Restoring soils and ecosystems made more sense to her than

Right: Maya Dutta organises the planting of micro forests in the city using the Miyawaki method.

regulating carbon emissions through mechanistic approaches. The idea of living systems intuitively resonated with her. She came across the non-profit Bio4Climate, a group dedicated to raising awareness of biodiversity and natural ecosystems as a climate solution. She worked as a volunteer for the group, reviewing the scientific literature and writing summaries of the latest findings. After only a few months as a volunteer, she was offered a full-time position with the organisation.

Soon after her arrival, the Bio4Climate team heard about Miyawaki forests. Finding the idea interesting, they decided to take on these urban forestry projects as a wonderful way of learning about ecosystem restoration on a small scale. Maya was chosen as project manager to carry it out.

The method was developed by Japanese botanist, Akira Miyawaki. In Japan, trees planted around temples, cemeteries and shrines are forbidden to be cut.

Akira Miyawaki believed that these trees are relics of indigenous trees once native to Japan, whereas most forests are now populated by trees, mainly conifers, imported for the logging industry. Unlike many tree planting initiatives today, which plant monocultures of non-native trees, Miyawaki believed that preserving natural forests with a diversity of native species was critical to establishing climate stability. In the 1970s he developed his method, supported by a contract he was awarded from a steel company to plant forests around their steelwork properties. Following that, he applied it to projects in Malaysia, India, Europe and the US. Miyawaki died in 2021 at the age of 93 but his work carries on, championed by organisations based in the many places he had worked and built relationships.

The method involves identifying a range of indigenous plant species to occupy different levels of canopy in a natural forest. There are the tallest trees which eventually come to dominate, accompanied by smaller, faster growing trees, shrubs and understorey vegetation. It is important to research the plants that were indigenous and not imported to the region. Next, seeds or seedlings are sourced or grown. Miyawaki acquired seeds from trees in sacred sites.

Before planting, the ground is prepared to ensure that the forest soil is rich enough for the plants to grow. A hole is dug in the soil about a metre deep; the earth is mixed with natural amendments to add organic matter, to help it hold water and stimulate microbial activity. The plants are placed closer together than is normally the case, anywhere from 3-7 plants per metre and in a random pattern, not a grid. The idea is that as they grow and compete for light, they will close the canopy, protecting the soil. In the meantime, the ground is covered with organic mulch material to keep it from drying out. The trees are tied to stakes for support and watered generously and frequently in the beginning.

These trees will grow more rapidly than trees in the forest because they lack competition and are exposed to plenty of light. The method accelerates the process of natural growth and succession, the result being a dense, maintenance-

Left: Maya is adding the final mulch to protect the newly planted trees. This microforest will cool, hold moisture and add significant biomass and biodiversity to an urban centre.

free forest in only 20-30 years. Without further human intervention, a natural process of succession will occur as the plants compete for sunlight.

Maya and her team consulted Miyawaki forest planters around the world along with indigenous peoples for almost a year before planting their first forest in the Cambridge area in September 2021. The plot covered 1219 square metres, on which 1400 saplings were planted. In November of 2022, they planted a second, smaller forest of 457 square metres elsewhere in the city. The climax trees are a mix of iconic New England maple and oaks, along with smaller species such as sumac, dogwood, witch hazel and plenty of shrubs and berries. Maya delights in witnessing the change of colours with the seasons, the progression of flowers attracting pollinators, as well as harvesting edible berries. It is an opportunity to involve school children, researchers and citizen scientists to document the evolution of a forest and the animal species that come to make their homes there.

For Maya and her colleagues, it is a balm for the soul to touch nature directly and share it with other nature hungry folks in the city. They plan to

continue building a resource for all, and are looking to work with businesses and learning institutions to plant more Miyawaki forests. This will involve compiling information about species native to a particular region, building repositories of seeds and native plants to supply future projects. Maya's major ambition is to restore the Northeast American seaboard to something closer to its original forested state, while addressing climate justice and giving more people a chance to learn about and access nature. She hopes that Miyawaki forests will be a part of the state's climate resilience plans and envisions planting them near health clinics and hospitals, schools and universities, and in underrepresented neighbourhoods.

Kahlil Baker: Carbon markets to plant trees and support communities

Kahlil Baker is a long-time tree planter as well as founder and CEO of Taking Root, an organisation which takes advantage of the funding from carbon markets to reforest degraded land by partnering with smallholder farmers. Carbon markets are one of the most developed financial opportunities today that channel funds into ecosystem restoration, and although they are not perfect, they help organisations like Taking Root to scale up their restoration efforts while improving the lives of farmers and their communities.

Kahlil grew up in Quebec, Canada, in a family of environmentalists living in a rural "intentional community", which is a communal group organised around a shared set of values. As far back as Kahill can remember, he has always felt a sense of mission to serve the environment and people. His first exposure to the global South occurred when he was in college. As part of a cultural exchange programme called North/South, he spent a month living with a family in rural Nicaragua. He developed a strong connection with the family and the country. This was the beginning of a long relationship. Twenty years later, he is still in touch and visits them almost every year.

Left: A Miyawaki forest after only two years of growth

Kahlil began his career as a tree planter, working summers for tree planting companies in Canada. Logging companies are obliged to fund the replanting of areas they have harvested. They hire silviculture companies to carry out intensive planting of tree seedling each spring. Participants are housed in a camp. They are paid by the number of seedlings they plant, so those who perfect their skills can earn a fair bit of money in a short time. Kahlil cultivated his skills for several years and made a lifestyle out of it. Once he had finished college, he would go tree planting for a few months each year and then travel the world for the rest of the year. Central America remains one of his favourite destinations. He continues to spend a lot of time in Costa Rica and Nicaragua.

Growing up, Kahlil had always thought of deforestation as a bad thing and those who deforest as bad people. But while he was in Nicaragua, he saw that people cut down trees to grow crops or graze cattle because there were few opportunities to make a living otherwise. Nicaragua is a poor country, recovering from years of civil war. Much of the forests have been cleared for agriculture. Kahlil wondered if he could find a way to combine his love for trees, travel and the people of Nicaragua into a reforestation

business. He wanted to find if there was a way to help people make a living while protecting forests.

His first attempt at a reforestation business with a couple of partners was the "Tree for Three" plan. They calculated that if they could raise three US dollars to plant a tree, they would plant one tree, pay one dollar to the landowner and invest one dollar in their business. However, it was difficult to raise significant amounts of money to pay smallholder farmers to grow trees. They managed to raise about $3000, far short of what was necessary to establish a business. Over the next few years, he formed other partnerships and tested other schemes, none of which proved fruitful. Kahlil remained determined despite losing several sets of business partners and other setbacks. He received two bits of advice during this time which changed his views. One was: if he wanted to start a business, why not go to university and learn how to run a business? The other suggestion was: why not sell carbon instead of reforestation? Awareness about climate change was growing and carbon markets were being touted as a climate mitigation strategy. In Kahlil's mind, carbon was only one of the many benefits that forests provide, not even the most important one, but if it could help him fund his project it seemed worth looking into.

Kahlil completed a BA in Economics, although he learned in the process that business and economics were not the same thing, but it ended up serving him well. He began researching the nascent carbon markets and found that there was a strong interest in tree planting projects. He incorporated Taking Root, the last iteration of his business efforts, as a Canadian non-profit in 2007, with the mission of establishing a sustainable community reforestation project in the global south. He began applying for the right to sell carbon credits. In 2010 Taking Root was certified by Plan Vivo, a foundation based in Scotland. Plan Vivo has developed a recognised methodology for certifying carbon sequestration. The foundation specialises in making their process easy for smallholder farmers and communities.

Left: Kahlil Baker and Elvin Castellon have been partners in the Communi-tree project in Nicaragua since 2007.

The concept was that Taking Root would sell carbon credits to companies around the world and partner with local community organisations in the tropics to access local farmers. Smallholder farmers would be given training and incentives for planting trees. Community organisations would act as intermediaries between Taking Root and local farmers to determine the needs of the community and to facilitate a collaborative reforestation plan, which would require not only planting the trees but ensuring their health in the years to come.

The next challenge for Kahlil was to identify community organisations in Nicaragua that could work with Taking Root. He and his colleague made a list of potential partner organisations, including a local NGO, Aprodein, which did many types of community development projects and they arranged to meet with its director, Elvin Castellon. On the way to the meeting Kahlil and his colleague got into a car accident. The car was wrecked, so they had to phone Elvin and inform him that they couldn't make it. Elvin kindly offered to pick them up and invited them to stay at his house. Kahlil and his friend were young, Elvin loved their playfulness and they all hit it off right away. They acted silly and played games but they still got around to talking business. They never bothered contacting other potential partners. Kahlil has worked with Elvin ever since, always staying at his home when he visits Nicaragua. The two men have become like brothers over the years. They have seen each other's children grow up.

While his business grew, Kahlil continued his education at the University of British Columbia, completing a Masters in Forestry and subsequently a Doctorate in Forest Economics, using his project with Taking Root as the subject of his thesis. Interest in carbon markets has grown and so has Taking Root. The demand for reforestation projects is greater than the capacity to supply them. Yet not all tree planting projects are of equal value or sustainability. A large percentage of the trees in planting initiatives around the world do not survive. The lack of incentives for caring for them is usually the cause.

Taking Root's success at keeping trees alive results from their approach to community support. The organisation takes the time, using local teams like the ones coordinated by Elvin, to understand the needs of the community and they propose reforestation solutions which will improve the lives of the smallholder farmers. Not all plantations are alike. If a farmer has grazing animals, the trees planted must complement grazing, by providing shade for the animals for example. The trees add new income streams for the farmers as well as improve the productivity of their existing operations. Taking Root has developed partnerships with coffee companies to help the farmers access markets for coffee grown under a tree canopy, known as "shade-grown" coffee. In some places early successional trees are harvested after a few years for timber and are replaced by saplings. Taking Root has helped to create markets for forest products which do not require cutting down all the trees, such as wooden ustensils and biochar. Carbon credits provide the additional benefit of continuing to pay for trees still growing.

The approach refined in Nicaragua is now the prototype for similar programmes elsewhere. Taking Root currently has pilot projects in fifteen countries, covering Africa, South America, Latin America and the Caribbean. They have developed a mobile phone app that enables smallholder farmers to upload impact data about the growth of their trees. By consolidating data from many small properties, Taking Root permits farmers to benefit from carbon payments. Big corporations buy carbon credits for large swaths of land and are not set up to tabulate data and payments for individual farms. Companies like Taking Root serve as the intermediary to create the bridge.

Carbon markets have come under heavy criticism. Large corporations sometimes use carbon sequestration projects to grab land and plant monocultures of commercially profitable trees such as oil palms. This takes the land away from local people and provides little benefit to them or the ecosystem. Some people believe that carbon offsets are simply a scheme to allow big companies to keep on with business as usual, as they can claim they are progressing towards carbon neutrality. Impact reporting may be dubious, and companies may claim benefits more than once or exaggerate the success of their initiative.

Despite these weaknesses in the system, which require continual oversight to prevent unethical behaviour, carbon markets continue to provide a significant source of funding for regeneration projects which might never have been realised without them.

John D. Liu: Ecosystem restoration camps and communities

This book started with John D. Liu and his Three Pillars of Life — carbon, water and biodiversity — and will end with his story. John has a vision of how to shift the way that humans live on the Earth through a grassroots movement around ecosystem restoration, and is building such communities all around the world.

He was born and grew up in the US, the son of a Scottish mother and a Chinese father. As a young man, John's father was in the Chinese nationalist army when he was badly wounded fighting in Burma. He was sent to London to recover. Afterwards, he planned to return to China but because of political strife there, decided to go to the United States instead. China's isolationist policy between 1956-1978 made it impossible for him to return to see his family. Part of his heart always remained in his country and he encouraged John to go to China as soon as he possibly could.

John studied television production and went to China in 1979, when he was 26. He helped to build production studios for an educational television station that was to broadcast Chinese language training programmes. In 1981, he was hired by CBS News as a production cameraman, helping to establish their bureau in Beijing. As a videographer, John covered the rise of China from poverty and isolation to a manufacturing capital. Developed countries opened factories in the country, profiting from cheap, skilled labour. The level of environmental pollution that resulted was unprecedented.

As the levels of pollution continued to rise, John felt the urgency to take action. He started the Environmental Education Media Project for China, collecting films on environmental issues and having them translated into Chinese. By gathering films, books and articles John was educating himself at the same time. Over a decade, he brought 1,000 films to China. He was also instrumental in the creation of the China Environment and Sustainable Development Reference and Research Center, which became a repository of vital environmental information.

At the time, the HIV/AIDS epidemic was spreading. John assisted in putting together the China – then United Nations - HIV/AIDS Information Centre. These efforts convinced him of the power of public awareness initiatives. He saw that if people understood an issue, they were less likely to be manipulated by policy makers, and more likely to act from their own understanding.

In 1993 he was invited by the World Bank to make a documentary about poverty reduction in China. The following year he was asked to document a baseline study for the Loess Plateau Watershed Rehabilitation Project in the upper and middle regions of the Yellow River. Participating in this massive landscape restoration project was life changing for John. The Loess Plateau is an area about the size of France, encompassing seven Chinese provinces. It had suffered from severe soil degradation, plunging inhabitants of the region into deep poverty. This massive restoration project transformed the region dramatically from arid and impoverished to lush and productive. John produced a documentary called *Lessons of the Loess Plateau* and then spent several years presenting his findings to audiences in China, Europe, Africa and the United States.

This experience helped John to understand that ecology is fundamental to human wellbeing. It took 10,000 years for humans to destroy the ecology of the Loess Plateau and only fifteen years to restore it. It proved that if such a large barren area could be transformed, then we could do the same all over the world. Environmental issues are the result of human mismanagement but ecology is an evolutionary process of succession and adaptation. He believed that if people could truly understand ecology, we could solve most of our problems on this Earth. We could stop toxic pollution, resolve the biodiversity crisis and fix the climate. John abandoned journalism to dedicate himself to speaking, making films and raising awareness about ecosystem restoration, which he calls **"the great work of our time"**.

With the Loess Plateau project, John had experienced on the ground. Until then, his experiences had been academic, working with high level institutions and organisations. He realised that there was a disconnection between science and the everyday lives of the people it aimed to address. Research agencies advanced the understanding of restoring degraded ecosystems, but this understanding only seemed to concern a small number of people. Public

Right: Mainsprings is one of the Ecosystem Restoration Communities located in Tanzania. Above is the site in 2012 before restoration began. Below, the same site in 2022.

policy was not being shaped as it should be. John felts that it was urgent to change the way we live before it was too late, and he was frustrated that it wasn't happening fast enough.

He began having dreams about a better life on Earth. The dreams persisted for months. Finally he decided to write an article, entitled "Earth Restoration Peace Camps", which portrayed an idyllic vision of people living in harmony together in a camp, working to restore the degraded landscape. It was published in 2016 in a permaculture magazine, then reprinted in other publications and circulated widely on social media. At the end of the article, John invited readers to pledge 10 euros per month to realise this vision. The response was overwhelmingly positive. Pledges poured in, more than enough to launch the first prototype Ecosystem Restoration Camp in 2017 on a five hectares property in Altiplano, Spain.

It signalled the beginning of the Ecosystem Restoration Camp movement. The funds are managed by two non-profit foundations; one is headquartered in the Netherlands and the other in the United States. As of 2023 there were 60 camps around the world with many more sites being considered. If the momentum continues there could soon be hundreds or thousands of them. Each is autonomous, but the foundations co-ordinate activities, share information and direct volunteers to the camps. In 2023, the name

was changed to Ecosystem Regeneration Communities (ERC) to reflect the reality that although there are many people who come to learn and stay for a short while, each camp is in service to a local community. The work that is done in each is relevant to the needs of the local people and the idea is to evolve a way of life which will continue for a long time. John joked that calling it a camp was the courtship phase of the relationship before deciding to move in together long term as a community.

John is the Chair of the ERC Advisory Council, but he has no financial interests and receives no salary. The ERC offers internships and training in ecosystem restoration. Online courses are also available. It has developed a framework using citizen science to monitor the impacts of restoration work, with guides tracking not only environmental changes, but also changes in the behaviour of members of the community and the surrounding area. Aside from the hands-on work of restoring degraded ecosystems, John wants the camps to become hubs for the creation of a new culture.

They function on participatory methods of governance and management based on theories of change management set out in Otto Scharmer's Theory

U and its predecessor, Participatory Rapid Appraisal. These are methods that groups can use to create systems together with input from all, to assess a changing situation and make collective decisions. This is a form of participatory democracy. Rather than chosen representatives making decisions for the people for a given term of governance, there is frequent consultation with those who will be affected by the decision. The governing committee gathers the inputs, formulates proposals and goes back to the stakeholders again for approval. The decisions taken with this process have more success with the community than those taken with a vote of majority plus one.

There are several pillars to this participatory, ecological stewardship culture. **Communal gardens** where master gardeners teach their skills to everyone is a central idea. They deliver food directly to **community kitchens**, which ensures not only that the origin of the food is known but that it entailed no unnecessary transport, packaging or food waste. All surplus food is processed by dehydration, fermentation, canning or other methods. The kitchen is also connected to the community's **cultural stage** where music, dance and theatre are performed and educational activities are held. A **Creator Space** is where tools and other equipment are shared. Members can build their own homes, furniture, and make repairs.

John is expanding this vision from the ERC to another type of cultural camp arising out of disaster zones. He and his partners, which include the Buckminster Fuller Foundation and the Schumacher Society, are trying to assemble teams to create transitional communities in disaster zones such as Syria and Turkey, countries which continue to struggle in the aftermath of devastating earthquakes or Somalia, which has been ripped apart by decades of conflict.

The Relief Center Design Labs will offer a combination of disaster relief and training for ecological restoration work along the lines of the ERC. Individuals with the necessary skills are being invited to meet online and advise. Like the ecosystem restoration communities, or the refugee camps,

Left: An evening gathering around a campfire at the Greenpop community in South Africa. Living together close to nature and united by a desire for social justice, equity and inclusion is a central feature of the Ecosystem Restoration Communities.

these camps can provide frontlines services in disaster zones and over time, repair their local ecosystems so they can grow food while regenerating the soil and meet their needs without degrading the land. They can learn how to live in cooperative self-organizing societies which value people and nature.

John believes that humanity is striving to take the next step in consciousness. We are sentient beings living on the only planet we know, one rich enough to sustain abundant life. We must create a culture based on respect for this life and protect its abundance and diversity. We cannot wait for others to bring about this change. Governments and the financial system will not lead the charge. Each of us must work with the people around us to stimulate change if we hope to realise this vision.

What can you do to help?

We have been warned again and again in multiple reports about climate change and the biodiversity crisis, and that the actions we take in the next decade will decide the future of humanity. The scale of change required is tremendous, and the timeframe is short, so we all need to be a part of that change. But how? You may as an individual feel powerless, but when enough people come together, much can be done, as John D. Liu discovered. **Margaret Mead said, "Never doubt that a small group of thoughtful, committed citizens can change the world; indeed, it's the only thing that ever has."** Each one of us has to find our own unique way to contribute according to personal circumstances and talents. But here are some things you can do now to get started.

Think indigenously

We have seen that much harm has been done by modern industrial culture because we have lost our connection to the earth. Indigenous cultures have a profound connection to place. Their traditional foods are local food, they understand the need to preserve the resources that their lives depend on.

Many of the people whose stories I've shared recounted that being close to the land changed something deep inside. They felt that the land was telling them what it needed.

The first thing we can all do is develop a relationship with the land where we live. Go outside, walk in the woods, become familiar with the changes of season. What is the natural biome where you live? How did the land look a 100 or 500 years ago? Do you know which watershed you are a part of? What is the source of your water? Answering these questions awakens your curiosity and sharpens awareness. Identify the flora and fauna, draw it, photograph it, collect wild edibles and medicinal herbs. Any activity which brings you into the present moment with nature is valuable.

Start or join a group to explore and protect the natural resources where you live. Are the waterways, the soil or forests under threat? It is great to balance a sense of mission which gives you a reason to be involved, with a sense of connection to your environment.

Wild your garden

We are accustomed to tidiness. When we build houses, we bulldoze the land and build on it. We plant lawns and perhaps a few shrubs and flowers. We do something similar to grow food, clear the land, plow and plant uniform rows of monocrops. Collectively, we need to realise that the biodiversity of a natural ecosystem is what provides us with clean air, water and fertile soil. We need to abandon the aesthetic of controlled, barren tidiness.

If you take an inventory of all the diversity you can observe in one square meter of land untended by humans, you will be amazed by how many different species are sharing that small space. By contrast, the biodiversity in most land that humans have tended is impoverished. The more people allow pockets of nature to grow wild in the midst of our communities, the more we will become accustomed to a natural aesthetic. Explain to others why it's good to foster native, biodiverse unruliness. It helps to sequester carbon, water and biodiversity which regulate climate. Your small area may not have a huge impact on the local climate but the more people you influence the more likely it is to add up to a cultural attitude change. Let's allow sections of our lots to grow wild, stop mowing vast lawns and allow natural habitats for birds, bees, amphibians and small mammals to co-exist with us.

Where does your food come from?

Most of us shop at a grocery store, giving little thought to where or how the food was produced. But food production has a heavy impact on climate and environment. A product may be part of the problem or part of the solution. Industrial agriculture produces tremendous volumes of cheap food and ships it great distances around the world. Conventional farms lose soil carbon, dehydrate the landscape and diminish biodiversity. Oceans are pillaged to provide us with seafood and rainforests are cut down to grow fodder for cattle. The food produced this way has a massive environmental price tag, which is not included in the price you pay.

You can play a big part in changing this by choosing food which is produced locally and regeneratively. There are local movements in many parts of the world; there are community gardens. When you grow some food in your yard, on your balcony or in a community garden, it gives you that connection to the earth. It makes you more aware of what it takes to produce it. Even if you only grow a small quantity, it makes you appreciate the farmers who provide you with the rest. Get to know your local farmers, visit their farm, talk with them about what sustainable practices they use and why.

Groups of people are forming coops to buy together, thereby reducing the need for packaging, and processed foods. Buy pasture-raised meat, which has helped to sequester carbon and water in perennial grasslands, rather than buy factory farmed meat. Buy food that regenerates the land.

Vote with your money

Whether you are well-off or less well-off, you can still help by spending your money in the right way. If you have a pension fund, discuss with your financial advisor or fund manager what your money is invested in and see if you can divest from companies which are harming the environment and invest more in impact funds. The financial products available are limited but growing. Although "greenwashing" is a problem, there are legitimate investment products driven by the desire to close the finance gap and solve the biodiversity crisis. The more people question the products in which they are invested and express interest in greening their portfolios, the quicker this market will grow.

If you have the money to support a cause, you'll find many worthy ones. Get involved, even in a marginal way, with groups rallying to rebuild our ecosystems, defend indigenous lands or build new social structures. Work a few months in an ecosystem restoration community or donate a small monthly sum to support a non-profit organisation. Feeling that you are a part of a movement and contributing both morally and financially helps build momentum and also gives you a sense of being a part of the solution.

Listen to your calling

All the people I talked about suffered from eco-anxiety, a feeling that they should do something, without knowing what. Each finally discovered a path to respond to this anxiety, a path that led to a project which eventually became their life's work. Each of us has unique skills and connections. There is work to be done in every part of the world, in every social niche. What are your skills and interests? Is there a way your unique talents can be of service? We often have a narrow idea about taking up a cause, considering ourselves unworthy of the task. Pay attention to people doing work you admire; just learning more may lead to an unsuspected opportunity.

Conclusion

The goal of this book is to help broaden the conception of what climate action means. The dominant narrative is about the reduction of greenhouse gas emissions. However, the IPCC reports clearly state that without the inclusion of carbon removals alongside emissions reductions, we will not succeed in time.

We deceive ourselves if we think the biodiversity crisis is distinct from the climate crisis, despite the massive impact that the destruction of ecosystems is having on climate regulation. Furthermore, we overlook the enormous mitigation potential that regenerating these ecosystems could have.

As Millan Millan pointed out, climate change has two legs: the greenhouse effect and the change in land use. The loss of functional ecosystems, or what we call the biodiversity crisis, drives climate change as much as GHG emissions. It causes a disequilibrium in Earth's climate regulating systems.

We have disrupted the carbon cycle, the water cycle, the nutrient cycle, and these losses contribute to climate instability. We must reduce the amount of greenhouse gasses we emit and at the same time turn down the surface heat of the earth by keeping it covered with cooling, living ecosystems. A larger percentage of cooled surface temperatures would lead to a net decrease of warming by replacing heat islands of cities and bare soils with vegetation which cools through evapotranspiration.

If we look only at emissions reduction as our strategy and ignore the immense value of natural ecosystems for cooling and regulating climate, we might make choices that will have terrible consequences. We have seen that these ecosystems have value not only as carbon storage but they regulate climate through the water cycle as well. Destroying them could exacerbate the climate extremes we are already experiencing from the disrupted water cycle such as extended droughts and extreme storms.

Carbon is one pillar supporting the cycle of life. We have lost 50% of the living biomass on Earth. This living biomass cycles carbon and water. It provides habitat for biodiverse life, which makes landscapes fertile and abundant. Life is a complex, interconnected system which regulates itself if we keep it in balance.

Plants have the unique ability to transform energy coming from the Sun. Through photosynthesis they transform carbon into biomass. Through evapotranspiration they evacuate heat from the Earth's surface and cause clouds to form. In both terrestrial and marine ecosystems, plants are the base of the food chain, converting the Sun's energy into a form that can be consumed by other forms of life. All the other elements needed to create life are cycled through this movement between earth and sky. Plants cycle hydrogen, carbon and nitrogen from the air and minerals from the soil. Animals transport seeds, spread nutrients and pollinate. Neither plants nor animals could survive without their microbial helpers, which break down and recycle every element that exists to keep the closed loop cycle going.

Inadvertently, we have disrupted the cycle of life and with it goes the ability to sustain both life and climate. We must now put the restoration of life sustaining ecosystems into our plans for addressing climate change. Cutting emissions is not enough. Even if we could achieve Net 0 by tomorrow, it would not be enough. Excess carbon that has already been absorbed by the ocean will continue to be released long after we stop adding more to the atmosphere. It will take a very long time to stabilise the balance of gasses in the atmosphere by reducing emissions alone. We don't have that much time.

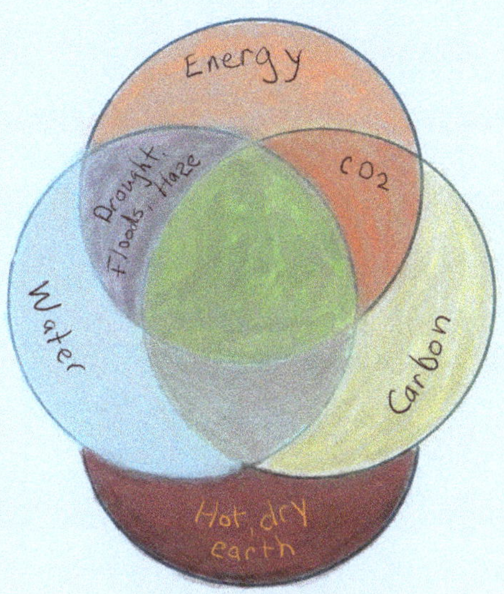

**Without living ecosystems
land dehydrates
releasing carbon, water + heat
into the atmosphere**

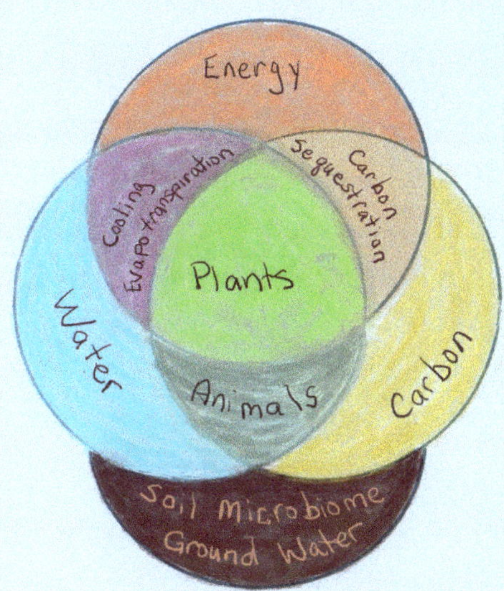

Water + Carbon = Life

**Life keeps planetary water +
carbon within balance
supporting Life**

This diagram shows how plants are at the centre of the cycle of life, cycling energy, driving the carbon and the water cycles. Without plants and their animal helpers, the carbon and water cycles cannot function to sustain life. The land becomes arid, the soil microbiome dies, the earth heats up and life as we know it cannot be sustained.

Another benefit of ecosystem restoration is that it changes things rapidly. If we could restore simultaneously large swaths of critical landscapes around the world, within a decade we would restore the small water cycle, bring back a cascade of life and sequester ever increasing amounts of carbon. And each subsequent decade would make the ecosystem more functional and resilient. All of this comes with a price tag which is ridiculously inexpensive compared to that of geo-engineering or disaster relief.

Nature is good at regeneration. We only need to let it do so. Millan Millan showed us how we enter into a vicious cycle of degeneration. The first misstep is removing too much vegetation. Once the vegetation is gone, the ground ceases to store water and the soil erodes. In areas where even sparse vegetation and soil remain, it will be quicker and easier to restore the water cycle and soil fertility. Let the plants grow and they will feed the soil and restart the water cycle. Once the soil is lost, however, it takes longer to rebuild it. Water must be captured in the landscape and planted with small hardy plants until the succession of life gains momentum.

There are three ways we, as humans, must go about to protect life on Earth. First, we must stop destroying the surviving mature, intact ecosystems. Second, we must restore the once healthy ecosystems which have been degraded. Third, we must create human managed ecosystems based on the ecological principles of natural ecosystems.

Two examples of the last approach are regenerative agriculture and green infrastructure in cities. There are countless models of climate friendly, regenerative agriculture which show us how to produce food while supporting the water cycles, providing habitat for biodiverse species and putting carbon into the ground. Some of these models are based on traditional knowledge, others are pioneered by groups of innovative farmers who blend traditional knowledge with modern innovations.

Green infrastructure adapts traditional ecological knowledge to our present context. With current weather extremes, the only way to provide any sort of sustainability is to design water resilience into our living and farming spaces. Ecosystems must be nested within cities. We can no longer afford to draw unlimited water from the ground. We must harvest water from rainfall, use plants to store it in the ground and cool our cities with green areas.

The EcoRestoration Alliance (ERA) is a new initiative which brings together ecosystem restoration practitioners, scientists, storytellers, engineers and educators who share the conviction that we must urgently create a groundswell movement to include the second leg of climate change (Millan

Millan's 'change in land use') into our climate action plans. Without the hopeful addition of ecosystem restoration — John D. Liu's "great work of our time" — we will never meet the challenge of climate change in time.

The Net 0 by 2050 objective, however important it is, will not on its own be enough for us to avoid catastrophic tipping points. The conviction of all the people interviewed for this book is that together can we scale up what we know works and move from a patchwork of ecosystem restoration projects scattered around the globe to a coherent, unified global strategy.

Almost anyone, almost anywhere can be a part of this movement. Any patch of land can contribute to regenerating a piece of its ecosystem. It can be as small as wilding in your back garden or as big as restarting the biotic pump to save the Amazon. Those with years of experience can be called upon to help newcomers get started. Mentors are eager to spread their knowledge.

The exciting thing about being a part of this movement is that it is within your reach and you see the results so quickly. Within a few years of ecosystem restoration work, you can see the environment changing. The water returns, the plants grow, the ground is more fertile, the air is cooler. You can see that restored land fares better than the surrounding areas when there is extreme weather.

No PhD is required to plant trees or construct a rainwater catchment from your roof or make compost. These gestures will give you pleasure and even more pleasure when you do them with others. You can find your way into this movement to rebuild the Pillars of Life. Twenty-billion acres of degraded land awaits to be restored. Imagine the result if we came together and brought it back to life.

Bibliography

Mark Nelson, *The Wasterwater Gardener: Preserving the planet one flush at a time*, Synergetic Press, June 2014.

Michal Kravcik, Jan Pokorny, Juraj Kohutiar, Martin Kováč, Eugen Tóth, *Water for the Recovery of Climate: A New Water Paradigm*, People and Water, 2007.

Walter Jehne, *Regenerate Earth*, Regenerate-earth.org, March 2018.

Ananstassia Makarieva, A.V. Nefiodov, Antonio Donato Nobre, Anya Rammig, The role of ecosystem transpiration in creating alternate moisture regimes by influencing atmospheric moisture convergence, Researchgate, May 2023, https://www.researchgate.net/publication/368643283_The_role_of_ecosystem_transpiration_in_creating_alternate_moisture_regimes_by_influencing_atmospheric_moisture_convergence

Judith Schwartz, *Forest modeling misses the water for the carbon:-QA with Antonio Nobre& Anastassia Makarieva*, Mongabay, February 2023.

David Ellison, Cindy E Morris, Bruno Locatelli, Douglas Sheil, Jane Cohen, Daniel Murdiyarso, Victoria Gutierrez, Meine Van Noordwijk, Irena F Creed, Jan Pokorny, David Gaveau, Dominick V Spracklen, Aida Bargués Tobella, Ulrik Ilstedt, Adriaan J Teuling, Solomon Gebreyohannis Gebrehiwot, David C Sands, Bart Muys, Bruno Verbist, Elaine Springgay, Yulia Sugandi, Caroline A Sullivan; *Trees, forests and water: Cool insights for a hot world*, Global Environmental Change; March 2017, https://www.sciencedirect.com/science/article/pii/S0959378017300134

Udaysankar S. Nair, Y Wu, J. Kala, T. J. Lyons, R. A. Pielke Sr., M. Hacker; *The role of land use change on the development and evolution of the west coast trough, convective clouds, and precipitation in southwest Australia*, Journal of Geophysical Research, April 2011.

Chloé Rebillard, « Le "paysan-chercheur" Félix Noblia invente l'agriculture sans pesticides et sans labour », *Reporterre*, February 2019.

Suzanne Simard, *Finding the Mother Tree: Discovering the wisdom of the forest*, Knopf, May 2021.

Sepp Holzer, Leila Dregger; *Desert or Paradise*, Leopold Stocker Publishing, 2011.

Kathy Dobie, An Amateur Rancher Brings the Wastelands of the Southwest Back to Life, *Oprah.com*, 2011.

James Lovelock, Gaia: A new look at life on Earth, Oxford Landmark Science, September 2000

Millan M. Millan, Climate/Water-Cycle Feedbacks in the Mediterranean: The role of Land-Use Changes and the Propagation of Perturbations at the Regional and Global Scale, *Researchgate*, January 2007.

Millán, Millán M., "Extreme hydrometeorological events and climate change predictions in Europe", *Journal of Hydrology*, Climatic change impact on water: Overcoming data and science gaps, October 2014.

International Peoples Assembly, Pakistan Underwater; Tricontinental, September 2022.

Michael Sheldrick, Hurricanes Highlight Climate Crisis as this Foundation Offers Hope, Forbes, October 2024.

Peter Bunyard, Winds and rain: the role of the biotic pump, MedCrave, December 2020.

Isabella Tree, *Wilding: The return of nature to a British farm*, MacMillan Publishers, March 2019.

What to do with this book once you don't want it anymore?

1. Give it a second life

Extend the life span of this book by giving it to someone or to a charity or by reselling it. A book with little damage can still be useful before being thrown away.

2. Recycle it well

If it has to be thrown away, put it in the recycling bin so that the paper it is made of can be reused. But be careful, it is important to remove recycling disruptors first:

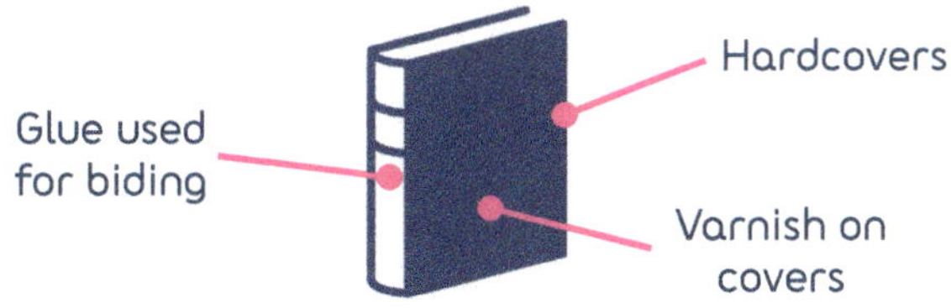

To do this, simply remove its cover and put it in the bin for packagings and all the pages left in the waste paper bin.

Legal deposit: December 2024